Technology and Industrial Transformation of China

Yanqing Jiang · Jiewei Gu

Technology and Industrial Transformation of China

Springer

Yanqing Jiang
School of Economics and Finance
Shanghai International Studies University
Shanghai, China

Jiewei Gu
School of Economics and Finance
Shanghai International Studies University
Shanghai, China

ISBN 978-981-19-7457-1 ISBN 978-981-19-7458-8 (eBook)
https://doi.org/10.1007/978-981-19-7458-8

This Springer imprint is published by the registered company Springer Nature Singapore Pte Ltd.
The registered company address is: 152 Beach Road, #21-01/04 Gateway East, Singapore 189721, Singapore

Introduction

Reform and opening-up plays an important role in China's technological development, economic growth and industrial transformation. The successful introduction of the market system has improved the efficiency of resource allocation in the social production process and promoted the input and renewal of physical capital, which has become an important factor in promoting economic growth in China. The success of China's economic growth has not only been related to the structural effect of institutional reform and the increase of investment, but also benefited from the technological progress embedded in material capital to a large extent. After the reform and opening up, China soon entered the catch-up period of economic development and industrial transformation, and its main task was to emulate the process experienced by industrialized countries and lay the foundation for industrialization. In fact, China has been developed in parallel in three main lines. The first is to reform the economic system, establish and improve the socialist market economy, promote enterprise reform, establish the dominant position of the enterprises in the market, and enhance the vitality of the enterprises. The second is to use large-scale resources and technology to build the foundations of energy, transportation, and communication and expand the production capacity of the processing industry. Third, China's accession to the World Trade Organization (WTO) is an opportunity to expand opening-up, participate in the international division of labor, exploit China's comparative advantages to a greater extent into the international industrial system and establish the foundation of China's manufacturing industry.

In fact, there are three main sources that may potentially lead to economic growth. They are, namely, from the perspective of the aggregate production function, capital input, labor input, as well as "total factor productivity" that is the very factor that results in output increase when capital and labor inputs remain unchanged. It is the improvement of production efficiency based on innovation and technological progress as captured by "total factor productivity" that is of top importance. A growth model that emphasizes only on the accumulation of material capital investment and labor input has its limits in that it would tend to lead to the result of a diminishing or even negative effect when the accumulation of production inputs reach a certain level. For instance, many industries have excess capacity, which is actually a manifestation of the unsustainability of such a model. On the other hand, in theory there would be no ceiling for technological progress and efficiency improvement. The richer the accumulation of knowledge can be, the higher the production efficiency it can induce. Such an innovation-based growth model then implies sustainability in economic growth and will be an inevitable choice for China to overcome the middle-income trap and move toward a highly developed country.

However, the transition to innovation-driven development is a process with a high threshold, and few countries and regions have successfully crossed this threshold. Innovation and technological progress are endogenous economic activities, which need to be driven by intrinsic motivation from individual market agents. Thus, the transformation of the economy requires deeper reforms. After the 18th National Congress of the Communist Party of China (CPC), the Central Committee soon confirmed the "new normal" of the transformation of China's economic development mode, from focusing on speed and quantity to efficiency and quality. China should give high priority to innovation, i.e., advocate entrepreneurship and innovation, and promote quality, efficiency and driving-force reform so as to raise labor productivity and total factor productivity. Enterprises should be the main participants of innovation, where innovation is the market behavior in which enterprises make risky decisions in order to obtain excess profits. Therefore, innovation investment requires more stringent environmental conditions than ordinary investment. Enterprises take their external environment as given and cannot change the external environment. Then, the majority of the market participants (the firms) tend to have convergent market behavior in response to the external environment. If just a few firms do not innovate, this might be due to their own decisions based on their own internal conditions, but if most firms do not innovate, we should then look into the external environment to see if there is a problem. In this sense, the issue of economic transformation is indeed the transformation of the development environment. In recent years, China's innovation at the firm level has developed rapidly. Significant progress has been made in both the investment of scientific research funds and the training of scientific and technological personnel. However, what cannot be ignored is that China's technological progress is characterized by various imbalances. China is still lagging behind developed countries in terms of independent innovation, material capital investment and human resource quality. In the years to come, China's technological development still has a long way to go.

China's adjustment, optimization and transformation of its industrial structure is a long-term strategic task. The world financial crisis that began in 2008 was an outbreak of violent fluctuations in the world business cycle and has greatly affected the global economic situation today. Many countries in the world are trying to adjust their economic structure and development mode in response to the crisis. China's response to the crisis was prompt and effective but by no means perfect. For example, the large-scale government investment represented by the four trillion investment fund in 2009 further magnified the defects of China's industrial system and structure. The slogan was to "stabilize growth and adjust structure," that is, stabilizing the GDP growth rate and adjusting the industrial structure. Evidence shows that there is still a great deal to be done to adjust the industrial structure in China, where the problems include large-scale overcapacity, long-term position at the low end of the industrial chain and value chain, weak international competitiveness in various industries, monotonous regional industrial structure, lagging development of the service industry, insufficient technological innovation, excessive dependence on natural resource endowments, dependence on exports, as well as the immature market system. It is obvious that the transformation of industrial structure plays an important role but nevertheless has become a somewhat obstacle to the rapid development of China's economy. It is therefore urgent to tackle the issue of industrial structure transformation.

Problems lying in the industrial structure in China, to a great extent, can be traced back to the industrial chain, supply chain and value chain. The industrial chain actually represents the close-knitted relationship between the supplies and demands of related industries, among which there currently exists structural problems. For example, overcapacity is the mismatch between the production supply and consumption demand. The supply chain and value chain are the relationships formed among industries and enterprises in the production process in terms of division of labor, coordination, technology matching and management rules. The low-end status of the industrial structure, plus circulation problems, excessive dependence on natural factor endowment and the export are all related to the governance of these chains. When the high-tech industry and other industries are isolated from each other, it will lead to the situation of long-term "enclave" contract manufacturing without technological upgrading. The dependence of the export on low-end products is actually shaped by the value chain controlled by multinational companies. Needless to say, there are reasons behind these problems, such as excessive administrative intervention by the government, decentralization of local finance, the protection of local markets and the so-called extensive development modes.

The industrial transformation of emerging powers is carried out under the conditions of economic globalization. The continuation of globalization has profoundly changed the development trajectory of the human society. The process of globalization will not be slowed down or stagnated by some countries adopting anti-globalization positions or policies. Anti-globalization does not mean the termination of globalization nor can it prevent further development of globalization. Since the outbreak of the financial crisis in 2008, countries around the world have adjusted their original economic structures and industrial systems. The so-called re-industrialization strategy started by the USA in 2009 had aimed to seek the rise of the American manufacturing industry with more advanced technologies and methods, occupy the high-end of the future manufacturing industry and bring the manufacturing industry back to the USA. In the process of the new globalization, the dominant power is bound to change; that is, all countries in the world have jointly participated in the new globalization. Apart from the USA, Japan and other developed economies continue to develop emerging industries, while BRICS countries are also exploring their own industrial development paths in the process of economic transformation. The third industrial revolution which has been widely disseminated since 2012 has initiated technological innovation in the future new-energy revolution, Internet revolution and intelligent manufacturing revolution. New industries have begun to develop rapidly, and a new industrial system is taking shape in the global landscape.

After summarizing relevant theories about industrial transformation, this book discusses the industrial chain from the perspective of globalization and further investigates the industrial development status of China. The industrial chain is the microscopic composition of the industrial structure, which is to say the industrial structure is the superposition of many industrial chains. The change and adjustment of the industrial chain will automatically promote the adjustment and transformation of the industrial structure. From the perspective of the industrial chain, it is not difficult to find that although China's manufacturing industry ranks first in the world in terms of volume it is highly dependent on imports for quality and technological development and is still at the low end of the global value chain. The original structure of global industrial chain has undergone shocking changes with the outbreak of COVID-19 in 2019. Before the outbreak of the pandemic, with the deepening of globalization, developed countries transferred the processing and manufacturing of final and semi-final products, parts and raw materials to developing countries, and earned high profits by utilizing capital and technology and forming industrial division among different countries. Meanwhile, according to the theory about horizontal labor division, multinational companies outsource the processing of some finished and semi-finished products in the industrial chain to global enterprises with good processing quality, high efficiency and low cost, forming an effective division of labor of the global industrial chain. The way of horizontal labor division that was formed was the mainstream feature of the industrial chain in the past. The outbreak of COVID-19 has halted global production, disrupted logistics and labor movement, which all have shaken the original industrial chain worldwide.

In the post-pandemic era, the question of how to restructure the global industrial chain has become the focus of this book. The "hyper-globalization" dominated by developed countries and characterized by the rapid flow of factors of production across the globe has receded. With the development of the digital economy and artificial intelligence, the increasing ability and willingness of emerging economies to participate in global governance, the industrialization process of developing countries and the objective needs of integrating into the global labor division will continue to support the development of globalization. The future globalization is a process of globalization where multiple forces compete and balance each other. The global industrial chain will reach a new steady state with the balance of technology, market, cost, competition, government and other factors. On the one hand, market, technology and cost factors will continue to drive the development of the global industrial chain. From the perspective of the market system, the growing domestic market of developing economies represented by China will continue to attract transnational corporations to form new regional agglomeration mode centering on terminal demand. On the other hand, from the perspective of the government and policy making, the influence of safety orientation in the global industrial layout has increased significantly. The impact of COVID-19 accelerated the trend of regionalization and localization of the global industrial chain that had emerged after the financial crisis.

How can China further realize its own industrial transformation in the process of global industrial chain reconstruction? One chapter of this book is devoted to this issue and proposes three ways to participate in the reconstruction of global value chain, namely embedding industrial transformation in the global value chain, constructing the national value chain and leading the regional value chain. According to industrial data published by WIOD, through calculating the VRCA index of five categories of China's high-tech industries, it can be concluded that raw chemical materials and chemical products manufacturing (C9), electrical equipment manufacturing (C13), electronic and optical equipment manufacturing (C14) should build their regional value chains while the pharmaceutical product manufacturing industry (C21) and transportation equipment manufacturing industry (C15) should construct a national value chain. The book also takes photovoltaic industry as an example to explore how China can further expand its competitive advantage in photovoltaic industry in the context of "Belt and Road" construction, and carry out relevant cooperation on technology and the production of spare components based on the actual development status of countries along the Belt and Road so as to achieve mutual benefit and win-win results.

With a new round of global scientific and technological revolution and industrial transformation poised for progress, it has become a new trend for current economic development to promote industrial upgrading and build a new driving force for economic growth with information and communication technology. China's information industry has also made remarkable achievements. However, it is undeniable that the fierce international competition environment has also brought some challenges to China's industrial transformation. Among them, there is a trend of lasting game between China and the USA around scientific and technological innovation and high-tech industry development. As the strength gap between the information industry of China and the USA is narrowing year by year, the scope of restrictive behavior of the USA is becoming wider and its intensity is also on the rise. In recent years, the USA has imposed direct restrictions or sanctions on a number of Chinese companies, seriously affecting their normal operations in the USA and even the entire overseas market. After the ZTE incident, the USA targeted Huawei, a Chinese brand with international influence in the information industry. The USA launched a series of attacks on Huawei starting in 2018. The restrictions led by USA at the market level have limited effectiveness, while the joint allies have cut off technology supply and supply chain against Huawei. Both of the two have a serious effect. In the long run, the USA has mainly adopted a combination of technological constraints and market compression. The US containment of Huawei reflects the impact of the rise of China's information industry on the leading role of the USA in making international rules. In the future, the competition pattern of technology and industry between China and the USA will become the focus and a long-lasing topic of the world. China should recognize the current competition pattern, gain an objective understanding of its own development path and effectively actualize technological innovation. Meanwhile, the importance of the international market should not be ignored. China should achieve faster and better industrial development from the two directions of technological development and expansion of the international market.

Based on China's national conditions and problems that exist in the process of development, the last chapter of the book puts forward China's plan for industrial transformation. The goal of a strong country can only be achieved through an innovation-driven industrial development path. In view of China's reality, the country needs to establish an overall awareness of industrial transformation and upgrading as well as a coordinated strategy to solve structural problems. China should strive to eliminate the blind spots in the choice of industrial development path and achieve orderly and coordinated development driven by innovation in the manufacturing industry. The overall consciousness is reflected in the efforts to build a modern industrial system in line with the direction of future development and the organic integration of various industries. The aim is to realize the high-tech transformation of traditional industries, from "heavy and thick" products to "light and small" products, and to use high and new technologies to transform traditional equipment and traditional production processes. In high-tech industries, breakthrough technological innovation should be used to facilitate industrial transformation. In addition, China needs to properly handle the relation between the government and the market, give full play to the vitality and innovation ability of market entities, and at the same time firmly adhere to the guiding role of the government policy on top of the allocation of resources as directed by the market so as to coordinate the interests of different entities. In the context of globalization, policy makers should properly understand the relationship between China's domestic and international situations. With this done, China will then truly have the ability to make full use of the "two markets" and "two resources" to upgrade and develop its manufacturing industry into a driving force to boost economic globalization and achieve win-win cooperation.

Contents

Chapter 1
Technological Transformation

This chapter discusses technological transformation and the changing of the world economic landscape. The development of technology not only provides conveniences for our social life, but also builds a bridge of communication for global information exchange and trade relations. In addition, the continuous development of the economy and the increase in trade demand also promote further development of technology. After the World War II, the world economy once dipped into a recession. If it had not been for the continuous progress of technology, the world economic order would not have been brought back again on track so promptly from the sluggish situation. In contrast, China has not kept up with previous technological revolutions, but instead the implementation of reform and opening up made China realize the importance of opening up. While actively introducing advanced technologies from developed countries in Europe and America, China also tried to innovate independently. However, decades of extensive economic development model brought many consequences to China's development. In the future, issues such as how China's economy can overcome the middle-income trap with technological development and how China can transform and upgrade its industries to adapt to the pace of economic development are the focus of this book.

1.1 Technological Transformation and the World Economy

1.1.1 Modern Technology

Before the technological revolution, the world economy was fragmented without a complete system. At first, modern natural science began to flourish during the Renaissance. From the sixteenth to the nineteenth century, natural science was a rigorous theoretical system based on experiments (or observations), with mathematics and logic as reasoning tools, which was separated from natural philosophy and became positive science. It is this kind of natural science in the form of positive study that

Y. Jiang and J. Gu, *Technology and Industrial Transformation of China*,
https://doi.org/10.1007/978-981-19-7458-8_1

gradually spawned technologies, which were then applied in the production process and became the earliest practice of applying technologies to production and human life.

The development of natural science since the turn of the twentieth century is incomparable from the sixteenth to the nineteenth century in terms of its depth, breadth, scale, and theoretical rigor. The revolution in physics during this period enabled science to penetrate into the microscopic realm. The discovery of X-rays and electrons has brought human understanding from the macroscopic world to the microscopic world. During this period, on the basis of quantum theory and relativity, many branches of physics, such as atomic physics, electron physics, and nuclear physics emerged, and various technologies developed based on these disciplines, which all have improved human's ability to understand and transform nature. The emergence and development of systems science is an important manifestation of natural sciences turning toward breadth in the twentieth century. In the 1940s and 1950s, information theory, cybernetics, and general system theory appeared successively. These theories focus on the structure of the system, the transmission of information inside and outside the system, and the control of information on the operation of the system from different aspects.

Notwithstanding the vigorous development of the natural science disciplines before the twentieth century, it is still the cause of only a few people and has not attracted the attention of the whole society. With its increasing scale, scientific research has gradually become a social undertaking with the support of national politics, economy and society. During this period, there was a phenomenon of the convergence of science and industry, which was highlighted by the emergence of industrial laboratories. The first industrial laboratory was founded in 1876 by the famous American inventor Edison. He broke through the "handicraft" research method of scientists in the past, and organized a group of specialized talents to work together on the same invention. The convergence of science and industry has greatly shortened the cycle of scientific inventions from experimentation to application. Before the twentieth century, some major scientific and technological achievements took a relatively long time from the invention (finding) of the principle to the application. The invention of the steam engine took 100 years, the steam locomotive 34 years, the telephone 56 years, and the electronic tube 31 years. Since the beginning of the twentieth century, the transformation of science and technology into direct productive forces has accelerated. The invention of radar took 31 years, the television 12 years, the transistor 5 years, and the laser only one year. The results of scientific experiments have entered direct application in the industry or the military.

1.1.2 Two Technological Revolutions

Based on the emergence of modern science and the overall development of science and technology, three technological revolutions took place from the eighteenth century to the mid-twentieth century, strongly promoting the development of social productive forces and the formation of the world economic system. Technological revolutions can be broadly categorized into two types, i.e. an overall technological revolution and the industrial technological revolution (steel, chemical, building materials, etc.). An overall technological revolution would affect all walks of life and fundamentally change the way of production during this period. Technological systems dominated by the steam power technology and mechanical technology replaced the use of simple tools operated by manpower or animal power. After that, in the second technological revolution, the technology system dominated by the electric technology and radio communication technology replaced the steam technology. The electronic technology-led system of the third technological revolution replaced the steam technology. In addition to the overall technological revolution, technological revolutions occurring in various industries or fields, such as the invention of trains, ships, and airplanes in the transportation industry, and the invention of electronic tubes, transistors, and integrated circuit in electronic technology have further economic growth.

1.1.2.1 The First Technological Revolution

In the seventeenth and eighteenth centuries, the completion of the primitive accumulation of capital created capital conditions for the future development of the bourgeoisie. During this period, the old technical level, especially the production method of the factory handicraft industry, could no longer meet the needs of social and economic development. Thus, the first technological revolution kicked off.

In 1705, Thomas Newcomen improved and invented a steam engine, which could already be used for pumping water out of the mine. However, due to the large volume of this steam engine, high coal consumption and low thermal efficiency, it has no economic value for promotion. In 1768, Watt invented the steam engine with a spilt condenser, which greatly improved the efficiency of the steam engine. But this steam engine is only suitable for production processes such as up-and-down and to-and-fro movement (e.g., water lifting and extraction out of mines), and has no general promotion value. It was not until 1784 that Watt invented the rotary steam engine, which could provide steam power for various working machines, thus the engine becoming the "universal prime mover" of universal promotion value. It can be said that the first technological revolution, started in 1695 (when the first steam engine appeared), lasted for about a century, and reached its peak in 1784 when Watt's "universal prime mover" emerged. Since then, steam engines have been widely used in industry, changing the entire production, and the technological revolution has turned into an industrial revolution. With the invention of the first rotary steam engine, steam power was applied to the spinning machine in 1785 and the loom in 1789. Since then, steam-powered technology has been widely adopted by the textile industry, and by 1824 the UK had 2,400 steam-powered textile machines. From 1800 to 1810, a total of 5,000 rotary steam engines were manufactured in the UK (an increase of 10 times in 10 years), which were widely used in papermaking, tanning, winemaking, grain processing, forestry and other sectors. By the middle of the nineteenth century, steam power had become the dominant force in mechanized production with water, wind, and animal power becoming the secondary force.

Due to the wide application of steam power, it has changed the appearance of traditional industrial sectors and promoted the rapid development of various industrial sectors. Among them, the textile, mining and metallurgical industries were particularly prominent. Most of the steam engines produced by steam engine factories were purchased by these three industries. Take a steam engine factory jointly established by Watt and Bolton as an example. Between 1775 and 1800, a total of 173 steam engines were manufactured, of which 93 were used in the textile industry, 52 in the mining industry, and 28 in the metallurgical industry. With the wide application of steam power, the textile industry, mining industry and metallurgical industry developed rapidly and became the three pillar industries at that time. Taking the UK as an example, from 1776 to 1780, when steam power was not widely used, the total annual export of textiles was 6.7 million pounds. After the introduction of steam power, from 1797 to 1800, the total annual export volume soared to 41.43 million pounds. In 1790, shortly after the introduction of steam power to the coal mining industry, the annual coal output was 7.6 million tons, while the annual coal output increased to 10 million tons in 1795. In the metallurgical industry, in 1788, when steam power was rarely used, the annual output of pig iron was 61,300 tons. In 1796, when more steam power was used, the annual output increased to 125,000 tons.

The widespread application of steam power became the symbol of the first technological revolution. The essence of this technological revolution was the replacement of the production method of workshop and handicraft industry by mechanized production. Working machines, steam engines and machine tools constituted the technical basis for mechanized production, all of which belong to the category of productivity. This productive force required the production relationship to be matched with it, that is, the modern factory system. The combination of machines and the modern factory system constituted the production mode of mechanized production. Compared with the production mode of workshop and handicraft industry, this production mode has obvious advantages (large scale, low cost, high labor productivity, good economic benefits, etc.) In the market competition, mechanized production gradually replaced the production mode of the handicraft industry and became the main body of the social and economic structure. In addition, this technological revolution not only has a positive impact on the economic and social development of the UK, but also expands worldwide. In the field of industrial economy, the United Kingdom, the United States, Germany and France have all achieved rapid development. Whether it is the output of raw materials for production or consumers, these four countries have come a long way. In terms of transportation, due to the introduction of steam engines into trains and ships, land and water transportations have developed greatly. The railways in the UK, the US, Germany and France have developed rapidly, as shown in Table 1.1.

With the development of the transportation industry, the international trade in this period also initially formed a scale. As the economy developed, various countries had surplus products that could be sold in exchange for their own scarce commodities. As a result, international trade has developed and reached a certain scale. World

Table 1.1 Overview of the economic development of the United Kingdom, the United States, Germany and France from 1850 to 1970

	Coal quantity (one million tons)				Pig iron production (one million tons)			
	UK	USA	Germany	France	UK	USA	Germany	France
1850	50	6.4	5.2	4.4	2.24	0.57	0.22	0.41
1870	112	70	26.4	13.3	6.1	1.7	1.4	1.2
	Cotton consumption (thousand tons)				Railway mileage (thousand kilometers)			
	UK	USA	Germany	France	UK	USA	Germany	France
1850	267	119	17.8	60	10.65	14.5	5.9	3.1
1870	488	224	81	69	25.5	85.2	18.7	17.9

Source Mendelsohn, *Theory and History of Economic Crises and Cycles*

trade was only 1.4 billion dollars in 1800, 7.2 billion dollars in 1840, and rose to 10.6 billion dollars in 1870, of which the UK, the cradle of the first technological revolution, accounted for 23.6%. To sum up, it is not difficult to find that, driven by the first technological revolution, in the 1850s and 1870s, the world's major capitalist countries have gradually transitioned from an agricultural society to an industrialized society.

1.1.2.2 The Second Technological Revolution

By the middle of the nineteenth century, with the development of the first technological revolution, major capitalist countries had established industrial systems of large-scale machine production to varying degrees and the production had developed rapidly. Capitalist countries actively developed the scale of enterprises, through the combination of new technology achievements and enterprise development they have obtained more and more economic benefits. By the 1860s and 1870s, the concentration of capital and production in the major capitalist countries had reached a fairly high level, and free competition had transitioned to monopolistic competition. The requirements for technological progress in this period were first manifested in the innovation of power technology. Although the steam engine achieved specialization and diversification in the early stage of its invention, the shortcomings of the steam engine were large in size, low in efficiency, and difficult to transmit power. In the 1820s and 1830s, the development of electromagnetism laid the theoretical foundation for electrical technology. The widespread application of electricity has become the main content of the second technological revolution.

The application of electric power includes two aspects, electric energy and electric information application. Electric energy has become a new power base to replace steam power in industrial production due to its convenient and efficient generation, transportation and distribution. Telecommunications connected various activities that were far away in space in time, greatly shortening the distance between people. The extensive application of electricity has changed the industrial production and even the whole social life, and has become the main content of the second technological revolution. In 1866, German technology entrepreneur Siemens successfully researched the first self-excited generator, which made the construction of large-capacity motors a reality. The emergence of Siemens motor for the application of electricity was equivalent to watt steam engine for steam power application. However, the application of electric power needed to solve more complex technical problems than the application of steam power. The application of electricity should not only solve the problems of power equipment such as generators and motors, but also solve the technical problems such as the construction of power stations and the long-distance transmission of electricity.

In 1882, American inventor Thomas Edison built the first power station, but it produced only 110 V of low-voltage electricity, which lost too much heat during transmission and travelled too short a distance. In order to solve this problem, French physicist Marcel Deprez developed a high-voltage transmission line, using the electricity generated by a three-horsepower turbine generator to transmit 1,500–2,000 V 57 km, driving a pump to form an artificial fountain. Deprez's mainstream electric transport was prone to safety issues, which has driven advances in AC high-voltage transmission technology. In the transformer (Stanley, 1885), AC motor (Tesla, 1888) on the basis of the German motor manufacturing company in 1889 (the main inventor for the Russian engineer Debrolisky) made the first three-phase generator. In 1891, under the auspices of Debrorowski, high-voltage transmission of 15–30 kV was used to transmit electricity to a distance of 175 km, with an efficiency of 70–80%, which was 4 times higher than that of DC transmission (22%). Since then, the effectiveness of high-voltage transmission technology has been recognized, and a solid foundation has been established for the large-scale promotion and electricity use.

With the establishment of the large-scale industrial production mode of machines and the wide application of electric energy, the production of industrial agriculture and domestic and international trade have developed rapidly. In the competition of the world market, the amount of information mastered becomes more and more important. Before the electromagnetic effect was discovered, people had sought various information contact or transmission methods, such as beacon towers, fast horse stations, etc. These methods were difficult to meet the requirements of long-distance and timely information transmission. The research on electromagnetism in the 1820s and 1960s provided scientific principles and technical ideas for the application of electrical information. The vigorous development of various telecommunication technologies ushered in a new era of telecommunication. In 1844, the first telegraph line was built and put into use. Since then, wired telegraphy has spread rapidly in the United States and around the world. In 1866 the submarine telegraph across the Atlantic was opened. In 1871 the telegraph service between Shanghai and London was constructed. The submarine telegraph cables that traverse all over the world connected all parts of the world and played a major role in the formation of the world market. In 1877 the telephone company established by Bell began the commercial production of telephones. In January 1878, the United States established the first telephone exchange, the Central Telephone Board. Since then, telephone service has been opened in Britain, France and many cities in Europe. Telegraph and telephone originally relied on wire lines for information transmission, radio communication could be achieved over a wider range of information transmission. Italian physicist Marconi and Russian physicist Popov successfully achieved bumpy communication, ushering in a new era of radio communication. People first heard radio broadcasts in 1901. Electromagnetic waves were used for navigation in 1911. In 1916, radio telephones were launched. In 1923, radio facsimile was realized. In 1926, black and white television was introduced. In the 1930s, an electromagnetic wave technology group was basically formed and widely used, and the electrical information industry rose rapidly. Around the widespread use of electricity, a series of new industrial sectors have emerged. In order to meet the requirements of power generation, transmission, distribution and use of electricity, generators, motors, transformers, and other electrical appliances manufacturing, installation, maintenance vigorously developed, formed a new electrical industry. In addition, emerging material industries such as conductive materials, magnetic materials and insulating materials for electrical appliance production continued to emerge.

The development of electric power technology has brought economic activities and even the whole social life into a new period. The rise of the electric motor has replaced the steam engine and become new power equipment in a large range, but the electric motor itself has certain limitations and cannot be used for various transport aircraft or machine tools. In order to overcome the shortcomings of steam engines and electric motors, the internal combustion engine came into being. In 1876, German engineer Otto finally invented the first practical internal combustion engine. However, Otto's internal combustion engine burned coal gas, which had defects such as low calorific value. Therefore, on the basis of Otto internal combustion engine, German engineer Daimler should use gasoline as fuel, and successfully manufactured the first gasoline internal combustion engine in 1883. In 1892, German engineer Diesel invented the diesel internal combustion engine. The invention of the different types of internal combustion engines mentioned above has greatly promoted the development of the transportation industry. Invention of the steam engine Steam trains appeared during the first technological revolution, but the invention of new heat engines led to a huge revolution in the transportation industry. Among them, the most prominent was the emergence of automobiles and airplanes. The rapid development and popularization of automobiles, airplanes and radio communications became the "three great inventions" in the early twentieth century. The development of the automobile industry has made the United States the country with the largest number of private cars in the world. The development of the aircraft industry marked the beginning of the air transportation industry and laid the foundation for the development of the air transportation industry after the 1930s.

1.1.3 The Initial Formation of the World Economic System

The second technological revolution not only changed social life, but also accelerated the formation of the world economic system. The economies of various countries scattered around the world, driven by the two technological revolutions, gradually become a whole that influences and depends on each other and complements each other, thus proclaiming the formation of the world economic system. From the perspective of technological revolution and industrial transformation, this chapter explains how the economies of various countries in the world constitute a whole through the international labor division, the world market, the information industry and the transportation industry.

1.1.3.1 The Major Capitalist Countries

The widespread application of electricity has driven technological breakthroughs in the transportation industry, and has stimulated the rapid development of industrial production in the European and American capitalist world. As shown in Table 1.2, between 1870 and 1913, when the second technological revolution occurred, the world industrial production more than quadrupled, achieving a growth rate that exceeded any previous period. However, during this period, more than 87% of the world's total industrial output was mainly contributed by the major capitalist countries in Europe and the United States that had undergone technological revolution, and the majority of Asian, African and Latin American countries that were technologically backward were still economically impoverished. It is not difficult to find that the discovery of the world economy in this period showed a preliminary phenomenon of inequality, and the main reason for this phenomenon was the gap in technological development.

Table 1.2 World industrial production index from 1700 to 1971

Year	The production index	Year	The production index
1701–1710	0.55	1894	42.2
1802–1812	3.18	1900	58.7
1820	4.16	1913	100
1840	7.4	1929	153.3
1860	14.7	1939	182.7
1870	19.5	1948	274
1880	26.9	1971	950
1890	41.1		

Source Rostow, *The World Economy: History and Prospects*

1.1.3.2 The Changing Nature of the World Market

Before the mid-eighteenth century, the world economy was not closely linked. Restricted by limited resources and products produced by traditional craftsmanship, international trade at that time mainly focused on precious specialties such as metal products, oriental spices and porcelain. These trade connections were mainly for the exchange of goods and did not form interdependent and interdependent economic links. The two technological revolutions promoted the development of the international labor division and changed the nature of the world market, so that the economies of various countries interacted and restricted each other and gradually formed a whole.

First, two technological revolutions promoted the establishment of an international division of labor system. After the machine industry replaced the production of the workshop and handicraft industry, the production scale of the industry expanded, and the sub-industry within the industry became more and more detailed. The expansion of the labor division has increased the variety of products and generally improved the quality of products, allowing large-scale industrial production to gradually go beyond the domestic base and go to the world. In the large-scale production mode, the raw materials needed for production cannot be self-sufficient in the country, and required to be imported from abroad. In addition, mass-produced products would have surplus products after meeting the daily needs of the country, which needed to be exported abroad. Of course, import and export were not completely free and were resisted, but the advanced technology and the production method of large-scale industry made the products of high quality and low price. The result of such technological development and product production was that foreign consumers could not resist the temptation and have to consume, and foreign producers had to produce to meet the product demand. In this situation, a new mode of international labor division that was compatible with machine production centers has emerged. The machine production center of the first technological revolution was in England, the cradle of the Industrial Revolution. After the second technological revolution, the machine production center was transferred to Germany, and the international division of labor system was further expanded and strengthened.

Second, the nature of world markets has changed. Before the technological revolution, the early world market was only commercial, which did not become an integral part of the expanded reproduction of industrial or agricultural capital. The nature of the world market, based on the international labor division, has changed. With the development of large-scale machine production, the quantity and variety of commodities entering the world market have increased substantially. In the past, the products traded on the world market were mainly some living materials or collectibles, which did not play an important role in domestic production. With the development of technology, various raw materials (cotton, wool, ore, coal, iron, etc.) and production equipment (machines, tools, etc.) have appeared in the products exchanged in the world market, that is to say, factors of production into the world market, which made both industrial and agricultural nation in the production process or economic activity closely link to the global economy. The global economic landscape was no

Table 1.3 Foreign investment of major capitalist countries from 1825 to 1913 (1 million dollars)

Year	1825	1840	1855	1870	1885	1990	1913
UK	500	750	2300	4900	7800	12,100	19,500
France	100	300	1000	2500	3300	5200	8600
Netherlands	300	200	300	500	1000	1100	1250
Germany	–	–	–	–	1900	4800	6700
USA	–	–	–	–	–	500	2500

Source Woodroof, *The Western Shock: A Study of Europe's Role in the World Economy*

longer the fragmented one in the past. The birth of the machine industry promoted the close connection of the world market and the operation of the world market was also inseparable from the production and consumption of the machine industry. This is the key point where the new world market in the era of machine mass production differs from the past world market in the era of traditional handicraft production.

In addition, the technological revolution facilitated the rise of capital markets. Except exporting industrial products and purchasing various raw materials to Asian, African and Latin American countries, industrialized countries also made a large amount of direct investment in these countries, set up factories, mines, shops and banks, and plundered the wealth and resources of these countries. Capital export became an important means by which the industrialized countries of Britain, France, Germany and the United States controlled the world economy. Table 1.3 shows the statistics of outbound investment of major capitalist countries from 1825 to 1913.

1.1.3.3 Telecommunications and Transport

The second technological revolution compared with the first technological revolution is the rapid development of the telecommunications industry. Telegraph, telephone, radio, fax and so on appeared one after another, connecting people living all over the world in terms of information. For economic activities, the role of the telegraph is particularly prominent. Telegrams sent messages by means of words (cryptography) that were timely and accurate. The laying of the European-Asian submarine cable constructed telegraph links between Europe and Asia. Trade partners or cooperative partners can conduct business negotiations in different places by telephone and settle trade settlement (telegraphic transfer) in different places by telegraph, even if they are far away from each other.

Another great progress of the second industrial revolution compared with the first one is the rapid development of the transportation industry. The invention of the internal combustion engine led to the advent of the automobile and the airplane and opened up new horizons in transportation. The invention of steam turbines and gas turbines led to the emergence of giant ocean-going ships, which greatly improved sea transportation capacity. According to European historical statistics, the registered tonnage of British steamships rose from 1113 thousand tons in 1870 to 11,273 thousand tons in 1913, an increase of nearly 10 times. French steamship tonnage also rose from 151 thousand tons in 1870 to 980 thousand tons in 1913. German steamship tonnage increased from 67 thousand tons to 2832 thousand tons. In addition, the development of technology has not only improved the maritime transportation capacity, but also greatly reduced the sea freight, which further promoted the development of international trade and closely linked the economies of various countries.

1.1.4 The Third Technological Revolution

The second technological revolution was mainly the discovery and application of electromagnetic effects. However, the electromagnetic phenomena and laws at that time stayed in the macro field, and the application of micro field was relatively little. The natural science revolution at the turn of the century broke the idea that atoms were the cornerstone of the universe, and human understanding entered the microscopic realm. With such conditions, electronic technology has gradually developed. The application of motor technology initially went through the vacuum tube stage, the transistor stage and the integrated circuit stage. In addition to these three major advances, a major achievement in the development of electronic technology in this period is the emergence of electronic computers. In 1946, Mauchler and Eckert of the United States developed the first electronic computer by replacing relays with electronic tubes and used it for military ballistic calculation. Since then, people have entered the computer era. In addition, the automation of the production process has undergone further innovations during this period. In 1948, Ford promoted the use of continuous automatic working machines in automobile production. In 1955, computers began to be used in power plants, oil refineries and other enterprises. By 1965, more than 600 factories in the world had realized the combination of production and computers, saving labor routes, reducing production costs, reducing the labor intensity of workers, and improving product quality.

In addition to the wide application of electronic technology, the emergence of the following technological inventions and related industries played an important role in promoting world economic development during this period. With the progress of geological survey technology and oil recovery technology, the petrochemical industry developed rapidly in this period. The world oil price was 9.82 dollars in 1870, but rose to 12.47 dollars in 1981. The rise of the petrochemical industry has shifted chemical feedstocks from coal to oil and gas. The combination of cheap raw materials, advanced technology and automated operations made the petrochemical industry the fastest growing industry in the world in the 1950s and 1960s, with an increase of 4.04 times from 1955 to 1971. With the deepening of people's research on the micro world, the nuclear energy technology developed on the basis of nuclear research constitutes an important aspect of the third technological revolution. From the perspective of technological development, the emergence of nuclear energy technology is a major technological revolution. However, due to the huge investment in atomic power plants and the higher cost of nuclear power than conventional power generation, nuclear energy technology has not been widely used. What's more, the development of space technology has promoted the development of the space industry. The first flight of the Croster jet took aviation to a new stage in 1941. During the World War II, due to the needs of the war, countries have developed rocket technology. In 1942, the German A4 large-scale rocket was successfully launched, achieving the goal of shooting 310 km away in 5 min. After World War II, space technology has also been developing in depth, which has opened up a new field for human future exploration, that is, the development and utilization of space resources.

Looking back at the background of the third technological revolution, we find that both world wars occurred during this period. The development of technology not only changes people's social life, promotes world economic relations, but also innovates the weapons of war. The new technological revolution also makes the weapons of war stronger and more lethal. But it is undeniable that the development of technology has still played a positive role in the recovery and development of the post-war world economy.

After World War II, the world formed a capitalist camp led by the United States and a socialist camp led by the Soviet Union. Due to the implementation of market economy, capitalist countries led by the United States absorbed the achievements of the third technological revolution and triggered the corresponding industrial revolution, which quickly recovered the economies of capitalist countries seriously damaged after the war and ushered in 20 years of sustainable development. In the 20 years between the two World wars, the average economic growth rate of major capitalist countries was 2.3%, and in the 20 years after the war, the average economic growth rate of these countries was 5.3%, which was faster than any other period in history and ushered in the "golden age" of capitalism. What is the driving factor behind such rapid development? Observing the GDP changes of capitalist countries after the war, we find that most of the economic growth of these countries depends on the improvement of labor productivity to achieve. From 1951 to 1987, labor productivity averaged 2.0% in the United States, 6.7% in Japan, 4.7% in Federal Germany, and 4.1% in France. In capitalist production relations, labor productivity is largely determined by scientific and technological factors, not by the enthusiasm of the workers themselves. Socialist countries led by the Soviet Union also achieved rapid economic development after World War II. The planned economy to a certain extent easily hinders the close integration of technology and economy. But this situation began to change with the implementation of economic reforms in socialist countries.

1.1.5 The Integration Trend

By integrating the changes of the world economic system, it is not difficult to find that with the continuous evolution of natural science since modern times, the technological revolution has undergone a transformation from steam power to electric power and then to electronic and nuclear power concentration. In this process, the world economy also gradually moves toward a whole from the original decentralization. With the acceleration of global transportation and information dissemination, the world economy has become closely linked. The development of technology is closely related to the economic trend, and the economic development will further promote the continuous transformation of technology. The integration of science and technology economy has become an inevitable historical development.

The emergence of modern natural science starts from experimental facts, builds a theoretical system and draws a series of specific conclusions based on abstract principles through mathematical or logical reasoning. If the conclusion is consistent with the facts after the practical test, the theoretical system is established. If it is not consistent with the facts, the theoretical basis needs to be revised until the test results are consistent with the facts. The scientific theory established by such an error correction mechanism is repeatable, verifiable and predictable, so that it can provide specific and effective guidance to production activities.

On the basis of modern science, steam technology as the core of the first technological revolution in the eighteenth century changed the operation mode of traditional manual farm, making the material production process become the application of technology in production and the science and technology become the productive forces acting on the economy and society. The development of manipulatory manufacture industry in Western Europe became the starting point of capitalism in history. With the germination and growth of capitalism in Europe, in order to maintain its economic interests and replace feudal politics politically, the emerging bourgeoisie set off the Renaissance movement and the Reformation movement and finally won the victory of the bourgeois revolution. With the continuous development of capitalism, the natural economy is gradually replaced by the commodity economy and the market economy.

Since the twentieth century, science and technology have developed extensively, forming a profound and broad network system, which has become the main source of the development of social productive forces and played an important role in the future economic development path and development direction. In particular, the new scientific and technological revolution since the 1970s, in the form of a whole related technological community, has continued to have a broad and profound impact on world economic development. Since the 1980s, social and economic system reform has been set off in the global scope. So far, the vast majority of countries in the world have chosen the market-oriented economic system, thus laying the foundation for the establishment of the global market economic system. This global market economic system is not a simple addition of the market economies of different countries, but a globalized economy formed by the organic connection and orderly operation of the economies of various countries through the world market, which put every enterprise directly in the face of fierce competition in the world market. The construction of the global information superhighway intensified this competition, which has deepened the thirst of enterprises for technological innovation and technological progress. In a word, the formation of science and technology as the first productive force and the world market economic system pushes forward the general combination of science and technology and economy to the internal and in-depth combination of science, technology and economy.

1.2 Technological Development and Economic Construction in China

The technological revolution has provided the world's major capitalist countries with an impetus for development, allowing them to enter the stage of foreign exchanges ahead of other countries. The development of trade and the primitive accumulation of capital provided market and financial support for these industrial countries. In contrast to China, we did not see China in the three technological revolutions. It is undeniable that in the initial stage of technological development, China did lag behind European and American countries, but since the reform and opening up, we have been constantly pursuing technological development and narrowing the gap with European and American countries in the process of realizing our own technological innovation.

With the guidance of the Communist Party of China, the Chinese people began a large-scale recovery and construction of the national economy in 1949. Before the founding of the People's Republic of China, due to political corruption, economic depression, and wars, China's science and technology did not receive due attention and development. The existing scientific and technological undertakings were incomplete, understaffed, clean economy and poor environment, falling far behind the world's advanced level. The Communist Party of China is soberly aware that the scientific and technological undertakings of new China must be restored and built in an organized and systematic manner. For this reason, the development of science and technology was brought under the strong leadership of the Party and the people's government, and efforts were made to change the declining situation of old China and bring science and technology onto the track of normal development. China has since opened a new era of scientific and technological development.

1.2.1 Project-Centered Industrial System in the First 10 Years

During the period of the first five-year Plan (1953–1957) and its subsequent years, new important industrial sectors were carried out with imported projects as the core. The "156 Projects" imported from the Soviet Union and Eastern European countries were the core projects of China's "First Five-Year Plan" construction, which included iron and steel complexes, non-ferrous metallurgy complexes, large coal mines, large oil refineries, heavy machinery factories, automobile factories and so on. At that time, almost all the new industrial sectors and important new products to be developed were supported by a number of complete introduction projects. On the basis of the introduction, we have formed the production capacity of dozens of important equipment such as duty vehicles, high-efficiency machine tools, automated blast furnaces and power generation equipment, and established an aviation industry capable of mass production of jet aircraft. The iron and steel industry has the manufacturing of alloy steel, seamless steel pipe, large-scale steel industry, non-ferrous metal industry has aluminum and other non-ferrous metal smelting and processing industry. The chemical industry has established divisions for the production of advanced dyes, industrial and aviation paint products. New radio and wire industries had been built, and telecommunications equipment, including receivers and transmitters and automatic telephone exchanges, had been made possible. By the end of 1956, China's machinery industry had produced thousands of new products, most of which were made after the patterns provided by the Soviet Union and eastern European countries. For example, among the more than 400 new products developed in 1957, more than 300 were imitations. During the first Five-Year Plan period, the proportion of means of production in the total import was as high as 92.4, 21% of the output value of China's heavy industry, and 55.9% of the import of machinery and equipment. Without such large-scale opening-up and introduction, China's modern industrial system would hardly have made such tremendous progress in a short period of time.

1.2.2 The 1960s and 1970s

In the early 1960s, the Soviet Union terminated most cooperation projects with China, and China began to turn to importing equipment and developing trade from other countries. In September 1962, China imported the first set of vinalon equipment from Japan. From 1963 to 1966, my country imported a number of equipment and technologies from Japan, the United Kingdom, France, Italy, the Federal Republic of Germany and other countries. In this period, the introduction of small and medium-sized complete sets of projects has promoted the development and improvement of the production capacity of China's petrochemical industry and other chemical industries. The production technology of the metallurgical industry has also been significantly improved, and the semiconductor and atomic energy have also developed rapidly.

In the early 1970s, with the resumption of exchanges between China and the United States, the establishment of diplomatic relations between China and Japan, and the participation of China in the United Nations, the external environment improved significantly, and China immediately deployed large-scale introduction of technology and equipment from Western industrialized countries. At that time, it was proposed to import 4.3 billion dollars of complete sets of equipment and single machines from abroad, that is, the "four-three" plan. After that, a number of projects were successively added, with a total value of 5.14 billion dollars. By 1977, the actual foreign contracted transactions reached 3.96 billion dollars. The introduced projects include 13 sets of complete sets of large chemical fertilizer equipment with an annual output of 300 thousand tons of synthetic ammonia and 480 thousand tons of urea using natural gas or light oil as raw materials, 4 sets of large chemical fiber equipment, 3 sets of large petrochemical equipment, 1 set of alkyl benzene plant, 3 large power stations, 43 sets of comprehensive coal mining units as well as turbine compressors, gas turbine manufacturing equipment and Bass engine projects. Almost all of these imported projects were weak links that were urgently needed in China's economic construction and production. Most of them were basic industrial projects, and most of their production technologies were relatively advanced in the world at that time.

1.2.3 The Opening Up

After the third Plenary Session of the eleventh Central Committee of the Communist Party of China (CPC), China fully implemented the opening up to the outside world. Despite the rapid improvement of economic development, import and export trade and foreign capital investment began to receive the attention of the government, and the autonomy of local governments opening to the outside world has been effectively mobilized. Economic development at that time was still in the preliminary stage both in scale and structure. Therefore, the policy of opening to the outside world has not been fully and effectively implemented, and the dividends of opening to the outside world have not been released to a greater extent. After 1992, China accelerated the pace of reform and opening up, proposed the establishment of a socialist market economy system, and brought it into line with the international market economy, and the policy of opening to the outside world was further implemented. This period is not only the acceleration of China's reform and opening up period, but also the high-speed growth period of China's economic development. At this stage, China put forward the development goal of establishing and perfecting the socialist market economic system and further released the huge vitality of the market. At the same time, China has put forward a series of preferential policies for opening up to encourage and support coastal enterprises to vigorously develop foreign processing trade, introduce foreign advanced production technology, and increase preferential policies for foreign investment. Thus, the pattern of "point, line and area" opening up from the coast to the border, along the river and extending to the interior has gradually taken shape.

In 2001, China joined the World Trade Organization to further expand its opening-up efforts. The opening-up of import and export trade and the introduction of foreign capital have been greatly enhanced, and the impetus for economic development has been rapidly enhanced. During this period, China's economic development has made remarkable achievements. After accession to the WTO, China accelerated the process of integration into the world economy, promoting the construction and perfection of China's socialist market economic system and national innovation system, thus broadening the road of China's economic development and national prosperity. China's accession to the WTO means a fundamental improvement in China's economic development environment, which is conducive to the realization of the third step of China's development strategy. With the continuous expansion of China's economic aggregate and the rapid improvement of its industrial technology level, it will be increasingly difficult to support the sustained and stable development of China's economy by relying on the technology imported from Europe and the United States in large quantities and cheaply, and the decisive role of China's local technological capability will become increasingly prominent.

In addition, China's entry into the WTO has fundamentally changed its investment environment. After the 1990s, multinational corporations began to transfer their R&D institutions overseas for the sake of localization of products, utilization of overseas R&D resources and establishment of global R&D network, which is reflected in the increase of overseas R&D branches, investment proportion and patent application. By the end of 2000, more than 100 multinational companies set up research and development institutions in China, of which more than 30 had achieved considerable scale, such as Motorola, General Motors, Volkswagen, Intel and Siemens. Research and development investment of multinational corporations is conducive to promoting the diffusion and spillover of China's technological progress and international advanced technology in domestic enterprises.

Technological development also brings new development opportunities to China's foreign trade. Since the 1960s, with the rapid development of world economy and technology, the status of international technology trade in world trade has risen rapidly. The total value of world technology trade was 2.5 billion dollars in the mid-1960s, increased to 11 billion dollars in the mid-1970s, exceeded 50 billion dollars in the mid-1980s, 110 billion dollars in the early 1990s and reached 400 billion dollars in 1996, accounting for 7.5% of the total value of world trade.

1.3 Difficulties in China's Economic Development

Since the reform and opening up, China's economy has achieved an unprecedented speed of development. Specifically, it can be divided into the following stages. The first stage was the low-income stage from 1978 to 1998. It took China 20 years to get out of the low-income stage, and the GNI per capita increased from 190 to 820 dollars. This is a decade of focusing on solving the problem of food and clothing. The fundamental driving force for economic growth is the reform of the socialist market economic system with the guidance of the basic policy of reform and opening up. The second stage is the lower middle-income stage from 1999 to 2009. At this stage, the input of resource factors is continuously increased, and economic growth mainly relies on factor investment as a driving force. The investment rate remained high, labor-intensive products "Made in China" spread all over the world, and foreign exchange reserves grew rapidly. The real estate became a pillar industry. The high investment, high cost and high pollution development model has also paid a heavy resource and environmental price, and the contradiction of unsustainable economic growth is prominent. The third stage is the upper middle-income stage from 2010 to the present. This stage is a critical period to overcome the middle-income trap. The decline in economic growth has switched to a medium-high speed and entered a new normal of economic development. In this new stage of development, it is necessary to transcend the mindset, abandon the traditional "path dependence", correctly understand the slowdown in growth rate, improve productivity, total factor productivity and potential growth rate as soon as possible, and replace the extensive past extensive development with intensive development. The driving force of growth has shifted from factor and investment-oriented to innovation-oriented.

China's problems are unique in the world. Thousands of years of ancient civilization and the accumulation of profound cultural traditions had a profound impact on all aspects of the economy and society. The great success since the reform and opening up has proved the correct choice of the road centered on economic construction, but there are also many shortcomings worth discussing. Since entering the lower middle-income stage in 1999, China has been in the middle-income stage for long time, and the economy has been in the middle-income stage for more than 30 years before falling into the middle-income trap. The reason why China will face the total income trap is inseparable from the impact of the world economic pattern.

Since the global financial crisis, although the world economic pattern has accelerated its evolution, the dominance of the United States and Europe has not fundamentally changed. Because of the first and second technological revolutions, the United Kingdom and the United States became the forerunners of the technological revolution. Relatively speaking, the pioneers often have more superior development space. After the industrial revolution, the United Kingdom, as the world's largest supplier of raw cotton and industrial products at that time, not only enjoyed the continuous supply of high-quality raw materials from all over the world, but also had the comparative advantage of continuously exporting industrial products to all parts of the world. After the second technological revolution, the United States took the lead and dominated the basic pattern of world economic development. However, the latecomers like China often faced a steep development environment, and were forced to accept the rules of the game formulated according to the interests of the pioneers, and suffered unjust and unreasonable exclusion and oppression.

China is a country with a large population in the world. Sufficient labor supply and cheap labor cost have promoted the rapid development of China's economy in the early stage of reform and opening up. Due to the planned birth policy and the change of people's conception of fertility, the birth rate in China has decreased seriously. According to the Statistics of the National Bureau of Statistics, the annual birth population in China has decreased from more than 26 million in the middle of the twentieth century to about 16 million at present, and the proportion of people aged 0 to 14 in the total population has decreased from 33.6% in 1982 to 16.5% in 2014, far lower than the world average of 27%. The number of children dropped from 340 million in 1982 to 220 million in 2014, a decline of 120 million in just 32 years. Demographic dividend is no longer, coupled with the aging of China ahead of schedule, old age before wealth has become a social concern. Compared with the aging process in developed countries, the aging process in China is not only characterized by a large absolute number of aging population and a fast-aging process, but also ahead of the modernization process. If we fail to transform and upgrade the economic structure timely, making development driven by innovation, and fully raising the level of human resources and labor competitiveness, per capita GDP growth will slow down more, which will put China squarely at risk of falling into the middle-income trap.

Since the reform and opening up, China's industrial structure has always been at the low level, and there is a significant gap with the world's industrial structure, which is highlighted by the high proportion of the primary and secondary industries and the low proportion of the tertiary industry compared with the developed countries. Since the world financial crisis in 2008, China's economy has been in a special stage of "three-phase superposition". The secondary industry is powerless, the tertiary industry is slump, the imbalance of industrial structure is prominent, and the transformation and upgrading are urgent. For the new normal of the economy, China is striving to optimize industrial layout and strengthen industrial upgrading. The primary industry should accelerate the transformation from traditional agriculture to modern agriculture. It is important to focus on solving the deleveraging, destocking, and capacity reduction of the secondary and tertiary industries while preventing economic stall. Meanwhile, we should strive to make domestic innovation in the equipment industry and increase value-added in the service industry.

In the middle-income stage, the resource scarcity and ecological environment constraints emphasized by economics are particularly prominent. For example, Japan was forced to turn its once proud heavy chemical industry into a processing and assembling industry due to the severe resource constraints brought about by the "oil crisis". In terms of resources and energy, due to its basic national conditions and resource endowment, China has objectively formed a pattern of excessive coal proportion. In addition, the extensive and extensive development model of the past 30 years has combined to synthesize multiple pressures on resources and the environment.

Overall, although China has not participated in the first two technological revolutions, it was not seen in the nascent world economic pattern. With the implementation of reform and opening-up, China has realized the importance of technological development and opening to the outside world. This book focuses on the discussion of China's technological development and industrial transformation, and explores how China can overcome the middle-income trap, realize technological development and use technology to promote industrial transformation in order to occupy a place in the changing tide of globalization.

Chapter 2
Technology and Economic Growth

This chapter introduces the interactive relationship between technology and economic growth in China. This chapter recapitulates the meaning and essence of technological progress. In this chapter we comb the evolution of theory concerning economic growth, expound the interactive relationship between technological development and economic growth from the perspective of the theoretical model, and discuss the mechanism through which technological progress has its impact on economic growth. We analyze the specific channels through which technological progress may potentially promote economic growth. In addition, in this chapter we also discuss the status quo and general trend of China's technology-supported development and sum up some problems in the development process.

2.1 Technological Progress

To explore the role of technological progress on economic growth, we must first accurately define the relevant concepts, especially the meaning and essential content of technological progress, along with its internal dynamic drives, which will have a direct or indirect impact on economic growth.

2.1.1 Implication of Technological Progress

With the continuous improvement of the natural sciences, technological progress plays a more and more critical role in economic growth. In fact, the scope of the concept of "technology" is quite broad. Macroscopically speaking, "technology" is a system of various knowledge, methods, means, processes, skills, know-how, tools, equipment and rules formed and developed in production activities, social practice and scientific experiments for the purposes of social production, social life, scientific

Y. Jiang and J. Gu, *Technology and Industrial Transformation of China*,
https://doi.org/10.1007/978-981-19-7458-8_2

experiment, scientific management and service provision. In our analysis, the definition of technology is constantly enriched and developed with the development of the times, and it has a historical and hierarchical implication. Technological progress refers to the production of more products with any combination of original factors of production than without technological progress, or the production of the same amount of products with fewer factors of production than before. Sometimes, the meaning of technological progress also includes the improvement of the quality of existing products, the emergence of new products and so on. Economists have done a lot of research on technological progress from both the micro- and macro-economic aspects. In microeconomic research, economists mainly explore the sources and causes of invention and technological innovation, or the speed at which new ideas and methods spread through the economy. At the macro level, economists mainly discuss the role of technological progress in the overall scope of the economy and the impact of the former on aggregate growth.

2.1.2 The Essence of Technological Progress

On the one hand, the purpose of technological progress is to eventually save labor and capital investment in the production process. Although the manifestations of technological progress may appear very different, yet from a historical view of technology evolution and development, the purpose of technological innovation and improvement as engaged in the production process is always to save workers' labor time and reduce labor intensity. As for the saving of capital, the role of technology itself is only the external manifestation of technological progress. Saving labor input in the production process not only leads to the change of production means, tools and labor objects, but also promotes the transformation from production experience to knowledge. Marx believes that "the material existence mode of labor acquisition machine requires natural force to replace human force and the conscious use of natural science to replace the rules derived from experience."

On the other hand, technological progress is essentially a reflection of the deepening division of labor. With the development of the division of labor in the production process, the production form of the whole society changes fundamentally. In this historical process, technology has achieved relative independence in existing form, so technological progress also has its own unique law. First of all, science that related to technology began to be independent from the production process. As a "completely objective production organism", the machine industry "decomposed each production process itself into various constituent elements, thus creating the completely modern science". The division of labor among various factors within technology has also gradually deepened. Scientific invention, knowledge accumulation, technological innovation and production application in technological progress are not only internally related, but also independent of each other, forming a unified whole. It is the deepening of the division of labor that makes the development of science and technology being from passive to active so that the technical basis of the social production process is always on innovation and constantly changes the existing production form and competition mode. Therefore, the technological foundation of modern industry is revolutionary. Changes in production inputs, industrial structure and even changes in social structure are all the result of technological progress in production.

The core of technological progress is the improvement of workers' ability and quality. In the process of social development, technological progress is reflected in the continuous decline of simple labor status, which has gradually become a weak element. Labor factors have been more and more linked with science and technology. Workers with certain scientific knowledge, production experience and labor skills are increasingly important to social production.

2.1.3 Technological Progress

2.1.3.1 The Subject of Technological Progress

Technological progress ultimately needs to be reflected in the process of social production. The choice of the subject of technological progress determines the success or failure of technological progress. Theoretically, enterprises are the main undertakers of social resource utilization and production activities, which should also be the pursuers of improving the level of technology. Technological progress in a broad sense includes technological innovation, which complicates the discussion on the subject of technological innovation in academia. From the perspective of innovation, enterprises are the decision-making subject, investment subject, research and development subject and income subject of technological innovation. At the micro level, enterprises are the main role of technological innovation. From the view of the enterprise inside, the entrepreneur's dominant and decisive power in technical factors directly determines the direction and process of enterprise technology, so the entrepreneur is also an important link in technological innovation. In addition to enterprises, more scholars believe that the subjects of technological progress also include research institutions, government agencies, markets and financial institutions. In general, the subjects participating in technological progress are diverse. Enterprises are the most directly related stakeholders in technological progress, enterprises are closer to the market, better understand the changes of market demand and can respond most directly to the changes of market demand. Therefore, enterprises have become the core subjects in the process of promoting technological progress.

Enterprises can also be further distinguished. There is also a relationship between technological progress and enterprise scale. Schumpeter believes that in terms of enterprise scale, large enterprises are more conducive to technological innovation than small enterprises. With the advantage of obtaining market profits, large enterprises have abundant capital for scientific research and development so that they can bear innovation risks. What's more, large enterprises also take a series of measures to prevent competitors from taking the lead in some innovation, so they are in a certain degree of initiative in the competition. In view of the risk of technological innovation, the expectation of monopoly profit of innovation activities and the ability of enterprises to bear innovation activities, large enterprises are more suitable to be regarded as the main subject of technological innovation. Therefore, it is obvious to draw a conclusion that the monopoly profits obtained by monopoly enterprises are an important source of large-scale enterprise innovation. Due to the external monopoly and competitive pressure, small and medium-sized enterprises have also become important subjects in technological progress. Monopoly enterprises already have ultra-high profit margins, so they will lose innovation activities that bear high risks and their sense of competition will be relatively weak. Small and medium-sized enterprises have better flexibility in enterprise organization and face a relatively single market. It is easier to seize opportunities, focus on R&D and implement innovation. When large-scale enterprises control the market, the process of technological innovation will inevitably slow down due to the lack of competitors. The coexistence and common competition of many smaller competitive enterprises will stimulate the technological innovation behavior of small and medium-sized enterprises so as to promote the technological progress of the whole department.

In fact, both large and small enterprises have certain advantages and disadvantages in technological innovation. From the perspective of the technological progress conditions, enterprises, whether large or small, should have the following basic characteristics: they have the need for technological innovation, that is, they can obtain high profits through the commercial application of technological inventions. In addition, they have the autonomy and decision-making power in technological progress and invention ability. With certain financial strength, they can apply and promote new technologies, bear risks so as to ensure the investment and subsequent application of technological innovation results.

2.1.3.2 Power of Technological Progress

In the academic research, there are also different emphases on the reasons for technological progress, which can be summarized as follows: market promotion theory, technology promotion theory, technological innovation integration theory and technology market interaction theory. The common feature of these innovation power theories is that they all emphasize the role of external factors making enterprise technological innovation. Hicks mentioned in *wage theory* that the change of relative factor price is an important reason for technological change and a large number of evidence shows that there is an internal correlation between the change trend of relative factor price and the saving factor tendency of technological change.

Combined with the discussion on the subject of technological progress in the previous part, the promotion of the external environment, especially the market competition mechanism on enterprises is an important reason to promote technological progress. Under different environmental conditions, the specific innovation power of most enterprises is not a single or fixed combination of dynamic factors, that is, they will show multi-level innovation power.

Fundamentally speaking, the pursuit of new profit characteristics is of decisive significance for enterprises to engage in technological innovation activities. The attribute of enterprises as economic entities determines that the internal driving factor of enterprises engaged in technological progress lies in their economic profitability. When enterprises introduce a "new combination" of production factors and conditions into the production system through technological innovation, it is easy to obtain potential profits through the process of market exchange. Under the condition of market economy, the benefit drove from innovation benefits can stimulate the awareness of technological progress and innovation spirit of internal personnel of enterprises, which forms the fundamental driving force of sustainable innovation of enterprises. Obviously, without this benefit driven, it is difficult for the external power sources of enterprise technological innovation, such as technology promotion and market demand to really play a substantive innovation driving role. External stakeholders (such as technology providers), entrepreneurs and enterprise employees are the main subjects undertaking different functions in technological progress. This spatial hierarchy determines the sustainable endogenous driving force of enterprise technological progress.

2.1.3.3 Realization Mechanism of Technological Progress

The realization mechanism of technological progress includes technological innovation and technological diffusion. The realization of technological progress is a process from part to the whole society. Promoted by scientific invention, technological progress is a starting point and the premise of the technological progress of the whole society. Technological diffusion is the condition of the technological progress for the whole society. The process of technological progress includes five aspects: basic research, applied research, research and development, commercialization of development achievements and diffusion of technological innovation achievements. From the perspective of social reproduction, technological progress should include the diffusion of technological innovation achievements, which is the unity of technological innovation and technological diffusion. From innovation to diffusion, diffusion drives the improvement of the productivity of the whole society, which reflects the realization mechanism and internal law of the social production process promoted by new knowledge and technology.

Technological innovation was first put forward by J. A. Schumpeter, an economist in *the theory of economic development* at the beginning of the twentieth century. He believes that innovation is the basis of economic development. Innovation introduces a new combination of production factors and production conditions into the production system, and lists a series of specific forms of innovation. After Schumpeter, many scholars have studied technological innovation. For example, through comprehensive arrangement, Muesser R. L. defined technological innovation as a meaningful discontinuous event characterized by its novelty of conception and successful realization. Drucker believed that the activities of technological innovation are to endow resources with new ability to create wealth. China has also clearly defined technological innovation. In the Decision on strengthening technological innovation, developing high technology and realizing industrialization of the CPC Central Committee and the State Council, technological innovation is defined as follows: enterprises use innovative knowledge, new technologies and processes, adopt new production methods and management modes, improve product quality, produce new products and provide new services to occupy the market and realize market value. It can be seen that there are two important links in technological innovation: one is to establish a new production process with higher efficiency and another is the realization of this new production process at the market level.

Technological innovation always needs to be realized in the market, so the diffusion of technology is crucial for a country or region. As for the concept of technology diffusion, Schumpeter defined it as the large-scale "imitation" of technological innovation. Some scholars regarded it as the "dissemination" or "transfer" of technological innovation, which means diffusion is the dissemination process of innovative technology among members of a social system over time through certain channels. "Diffusion" is the process of spreading innovation among a certain social member through a specific channel over a period of time. Chinese scholars Kuang and Tang (2006) simply defined technology diffusion as "the process of technology diffusion and adoption among potential users through certain channels. Generally, it refers to the new application of technology that brings expected economic benefits to new users. The process of technology diffusion includes not only the conscious technology transfer, but also the unconscious technology spillover of the technology subject". This book tends to define technology diffusion as a process in which new technologies are disseminated among potential users through certain channels, adopted by new users and bring expected economic benefits.

From the perspective of economics, technology diffusion is the source of social benefits of innovation and the economic benefits of a new technology mainly come from its diffusion. Only through the diffusion and large-scale application of new technologies in the society can the whole society achieve more economic benefits. The emergence of new technologies will form a huge social effect in the society. Those enterprises that have not adopted new technologies will imitate one after another and obtain excess profits on schedule so as to improve the economic and technological level of the whole society.

From the perspective of the ways of technology diffusion, this process includes not only the conscious technology transfer and technical transformation, but also the unconscious technology spillover of the subject. Generally speaking, technology flow between countries is technology transfer and technology flow within a country is technical transformation. Both technology transfer and technical transformation constitute the diffusion process of technological innovation. Technology spillover is different from the form of technology transfer and technical transformation. Technology spillover has a more unconscious behavior of the technology subject. In terms of actors, there is generally only one transferee of technology transfer and technical transformation. Technology spillover generally has multiple or even many recipients of new technologies and most of them are potential recipients, whose object is not clear.

2.2 Technology and Economic Growth

There is no doubt about the role of technological progress in promoting economic development. Scientific and technological innovation is the primary development force. Every leap in productivity in human history is inseparable from the progress and innovation of science and technology. The rapid development of technology and its transformation to modern productivity have increasingly become the most active factor and the main driving force in modern productivity. Technological progress plays a decisive role in a country's economic and social development. In particular, the first and second industrial revolution and technological revolution have promoted the western economy and the entire national strength, which has made the whole world realize the great power of scientific and technological innovation for economic growth. This is mainly because technological innovation generates a lot of surplus value and the leap in productivity helps the economy far more than the investment in research. The research on technological progress as an influencing factor of economic growth originated in western countries and has been improved and developed in these countries.

2.2.1 Classical Economic Growth Theory

Classical economic growth theory occupies an important position in economic growth theory and lays the basic context for the study of economic growth. However, the impact of scientific and technological progress on economic growth has been neglected in the study of classical economics.

Classical economic growth theory mainly focuses on labor and capital, emphasizes the role of physical capital, takes the scale of capital stock, especially the speed of capital accumulation as the core factor affecting economic development and develops the Harrod-Domar model as the core idea. The model is based on Keynesian theory and the specific assumptions are as follows: a country's savings can be effectively transformed into investment and the country has sufficient absorptive capacity for foreign capital transfer. If the society produces only one product, this product can be either consumer goods or investment goods and the only two production factors in production process are labor and capital. Moreover, the two elements can't replace each other, the capital-output ratio remains unchanged, the technical state is established, which means there is no technological progress and capital depreciation is not considered. Through formula reasoning, it is found that economic growth rate is a function of savings rate and capital output rate. We can see Eq. 2.1 that the growth rate increases with the increase of savings rate and decreases with the expansion of capital-output ratio.

$$\Delta Y/Y = s \times \Delta Y/\Delta K \tag{2.1}$$

In the equation, $\Delta Y/Y$ represents the economic growth rate, s represents the savings rate, $\Delta Y/\Delta K$ represents the output that can be increased for each additional unit of capital, that is, the use efficiency of capital (investment). The model expresses the role of capital transfer in economic development. Promoting economic growth can be achieved by increasing investment (savings rate). But this model does not involve the role of technological progress in economic growth and it also negates the substitutability of production factors. In current studies on economic growth, this model is often used to analyze long-term economic problems in developing countries and deal with the relationship between investment, savings and economic growth.

However, with the in-depth exploration of the theory by scholars, the technical factors have not been completely ignored. They attributed the scientific and technological factors to other factors rather than study them as a single factor. In the book *A Treatise of Taxes & Contributions*, Willian Petty mentioned that "land is the mother of wealth and labor is the father and active element of wealth", he attributed the source of commodity value to labor and regarded the improvement of labor productivity as the main factor to promote the growth of national wealth. Willian believes that wealth growth first promotes the improvement of labor productivity which depends on the level and scale of division of labor. The finer the division of labor, the higher the labor productivity, the lower the product cost and the more of profit.

What's more, the improvement of labor productivity also depends on technological progress. Adam Smith, the founder of the classical economic system, put forward in the *wealth of nations* that "the greatest improvement in labor productivity, proficiency, skills and judgment in the use of labor are the result of division of labor". Division of labor originates from exchange capacity. Therefore, the degree of division of labor is always limited by the size of exchange capacity, that is, by the wide and narrow market. Adam comprehensively researched the problem of capitalist economic growth in theory and practice. He not only analyzed the role of the three basic production factors of labor, land and capital in economic growth, but also discussed the problems of technological progress and improvement of labor productivity. Smith believes that the progress of production technology can reduce production costs and improve labor productivity, so as to increase national wealth. In industry, with the increase of labor and capital investment, production is increasing returns to scale. The existence of various laws of increasing returns is the result of production technology progress and the labor division development. Therefore, Smith attributed the growth of national wealth to the development of division of labor and capital accumulation, but the most fundamental reason for promoting economic development is the deepening and evolution of social labor division. Smith's theory of division of labor also confirms the role of technological progress in promoting production development.

Recalling the discussion of classical economists on economic growth, economists turned from the field of circulation to the field of production, searching for the source of wealth growth. The capital determinism that capital investment determines economic growth has always occupied a dominant position in the theory of economic growth, although some economists mentioned the role of technological progress in promoting economic growth. However, the understanding of the role of technological progress mostly remains in the stage of qualitative analysis.

2.2.2 Marxism

Marx believes that technology is the productive force. In *the logic of Capital*, he emphasizes that the expansion of production can not only rely on the increase of capital increment, but also play a role through technological progress. Marx pointed out that capital accumulation was not the only source of expanding reproduction. The main reasons for the expansion of production year by year are as follows. The capital invested in production is increasing and the efficiency of capital use continues to improve. Labor productivity is also determined by a variety of circumstances, including the average proficiency of workers, the level of scientific development, its application in technology and its combination with the production process, the scale and efficiency of means of production and natural conditions. In addition, Marx also put forward the famous thesis that "science is included in productivity", which can be transformed into productivity only through the bridge of technology. Therefore, the development level of productive forces is determined by the development degree of science and technology and on the basis of the development degree of science and technology. In turn, the development of productive forces promotes the emergence and development of science and technology. At the same time, social economic system has a strong restrictive effect on science and technology, that is, science and technology, social system and economic development together constitute an interdependent and mutually promoting community. However, Marx's expanded reproduction model does not include technological progress as an influencing factor in the model, but only as the embodiment of intermediary effect.

Based on the above, Marx's basic views on the relationship between technology and economic growth can be summarized as follows: first, in the specific process of labor and socialized mass production, the change of output must correspond to the corresponding change of various input factors. Second, for the sustained economic growth, the input growth of production factors is limited. It is inevitable to improve the output efficiency of input factors, so as to have more output under the condition of the same amount or a small amount of input. This goal can only be achieved through technological progress. Third, only technological progress can change the combination mode of a certain number of production factors and enhance their use value. In addition, the role of technology in various factors of production can also improve the level of production equipment, the quality of workers and the level of management, so as to promote the improvement of productivity.

2.2.3 Neoclassical Theory of Economic Growth

The research on economic growth theory can be divided into two periods: the first period was the neoclassical economic period, from the Harrod-Domar model in 1939 to the end of the 1960s, which was the first golden age of growth theory research. During this period, Solow and Denison successively proposed their own economic growth theories and models, which decomposed and measured the contributions of various factors of production (capital, labor, intermediate income, etc.) and technological progress to economic growth, pushing the research on the role of technological progress in economic growth to more practical aspects. In the late 1950s, aiming at the defects of the Harrod-Domar model, economists represented by Solow, Swan and others established the neoclassical economic growth theory (Solow 1957), which strengthened the focus on the factors of technological progress and gradually evolved into an important factor in modern economic growth. The research in this period has made great progress both theoretically and empirically. The main contribution was that the factors of technological progress were incorporated into the economic model as exogenous variables.

For example, Solow model took the factor of technological progress as the most significant and contributing factor of economic growth independently for the first time. On the basis of in-depth study of Harrod-Domar model, Solow broke through the model and put forward his own economic growth model by relaxing the assumption that capital and labor are irreplaceable, thus stabilizing the conditions of stable economic growth. Its production function is expressed as $Y = F[A(t)K, L]$. The corresponding growth rate is $\Delta Y/Y = \alpha \Delta K/K + (1-\alpha)\Delta L/L + \Delta A/A$. It shows that the growth of output is equal to the weighted average of capital and labor growth plus the growth rate due to technological progress. $\Delta A/A$ is often referred to as "total factor productivity" (TFP), which is referred to as "capital augmenting" technological progress because it results in an increase in output that has the same effect in the capital stock. Using the total factor productivity analysis method to test the model, the contribution of capital and labor input to economic growth is about 12.5% and most of the economic growth is not brought by these two inputs, but by an exogenous which is interpreted as the "residual value" of technological progress. Thus, Solow attributed the sources of economic growth to two factors, one was the "growth effect" generated by the increase in the number of factors, while the other was the "level effect" economic growth brought about by the improvement in the technical level of factors.

Thus, the new classical economic growth theory broke through the long-term economic growth theory's dominant point of view by saying that capital accumulation is a decisive factor of economic growth. This theory put forward the view that "technological progress has an important contribution to economic growth" for the first time and pointed out the decisive role of technological progress in economic growth in both theoretical and empirical research, which has greatly promoted the development of economic growth theory. However, the theory still has limitations. It only takes technological progress as an exogenous variable into the model, that is, it excludes the role of technological progress on long-term economic growth. At the same time, the neglect of the government's economic policies and too strict setting in the theory are not consistent with the actual situation of social and economic development.

2.2.4 New Economic Growth Theory

Since the mid-1980s, on the basis of rethinking the neoclassical growth theory, economists represented by Romer, Rebelo, Barro, Beeker and others have researched the fact of huge contrast between the economic growth of developed and developing countries at that time and published a series of academic papers on "endogenous technological progress". Romer's "Increasing Returns and Long-term Growth" and Lucas's "On the Mechanism of Economic Development" have opened up a new way for economic growth theory, which explains how technological progress affects economic growth and expands the research ideas in this field.

In the endogenous model of technological progress, Romer makes it clear that technological progress is an important factor of production that affects economic growth. The model emphasizes the importance of knowledge stock and overall technological progress comes from investment in physical capital. Romer used cumulative investment in production to represent the accumulation of knowledge, internalizing the technological level. As a special input in the production process, knowledge differs from ordinary factor input (capital and labor) in that knowledge has increasing marginal productivity and can improve the return on investment, thus promoting capital investment. In turn, capital investment can stimulate the generation of new knowledge, forming a virtuous cycle of capital investment and knowledge accumulation. Romer's conclusion from this model is that technological progress can increase the return on investment, which in turn leads to the accumulation of knowledge and speeds up the process of technological progress. In such positive and negative feedbacks, a virtuous cycle of growth appears in the economic system, thus achieving a stable long-term increase in economic growth rate. In view of the above model, Romer also optimized it, taking technological progress as a factor affecting economic growth, which should meet the corresponding theoretical premise: technological progress is the core of economic growth and endogenous, knowledge is a special public good, which can be reused infinitely without additional cost. In addition, knowledge is non-competitive and non-exclusive. At the same time, Romer divided the factors of production into four categories: capital, unskilled labor force, human capital and new ideas and summarized technological progress as the expression form of the latter two in economic growth. According to this, Romer divided the economic life into three sectors: R&D sector, intermediate product sector and final product sector, emphasizing that R&D sector is the source of economic growth. Among them, R&D department uses human capital and existing knowledge stock to produce new knowledge and its product form is used to produce the design of new capital goods. Romer's model takes the accumulated capital in production to represent the knowledge level at that time and internalizes technological progress. This model is also called knowledge accumulation model.

On the basis of Romer's research, aiming at the two major defects of the model, namely the diffusivity and the correlation between economic growth and capital accumulation, Lucas proposed the Arrow type specialized human capital accumulation model, that is, human capital accumulation can be obtained through formal education in schools. Furthermore, the "internal effect" of human capital obtained through school education and the "external effect" of human capital produced by learning while working. Thus, the differences in economic growth among countries can be well explained. The higher the level of human capital is, the richer the total human capital is, the higher the economic growth rate will be. However, in some backward countries, due to the low level of human capital, there is not enough human capital to invest in R&D activities, these countries will be locked in the "low-income trap", or even no economic growth.

In the new economic growth theory, the research on the effect of technological progress on economic growth has strong theoretical and practical significance. This theory makes up for the deficiency of knowledge accumulation and theoretical analysis of technological progress in traditional western economic theories and discusses that technology is the decisive factor of economic growth from the theoretical perspective, which deeply analyzes the mechanism of scientific and technological progress on economic growth through the establishment of relevant models. According to this theory, as a system variable technological progress is the most active central element in the current world economic system. In essence, economic growth is the synergistic result of many elements of capital, labor and technology in the system. In order to obtain high-quality economic growth, it is necessary to give full play to the role of technological progress factors on the basis of resource input. The theory not only explains the facts of economic growth well, but also provides effective theoretical support for governments to formulate long-term economic growth policies.

2.3 Economic Growth Determined by Technological Progress

The greatest effect of technological progress is to increase labor productivity and thus promote economic growth. Technological progress comes from the production of all links of knowledge on progress and innovation activities throughout the production process, technical advances both in improving and upgrading machinery, equipment, tools, modified material resources quality characteristics, etc. It also comes from the scientific research personnel, management personnel and the improvement of the quality of workers and intelligent, the improvement and innovation of the principles and methods of production decision-making, organization and management.

These factors will promote the improvement of labor productivity, thus promoting the process of economic growth.

In addition, the decisive effect of technological progress on economic growth also comes from the cumulative effect of technological knowledge. Technological progress has a cumulative effect, which can greatly increase the stock of technology and promote the rapid economic growth. The new economic growth theory also regards the mixture of theoretical knowledge and practical economy as the existence form of technology, and holds that technological progress is the increase of technological stock. As the marginal revenue of the accumulable elements remains unchanged, while the non-accumulable elements (mainly labor elements) remain unchanged, the economy will show increasing returns to scale. In other words, the progress of technological knowledge is cumulative. As the marginal income of this accumulable factor of production remains unchanged, it overcomes the problem of diminishing returns in the process of material capital input and can rapidly promote economic growth.

2.3.1 Technology as the Core Driving Force for Economic Growth

In the long history of social development, the efficiency of global economic growth was in low level for a long time. Until the 1860s, the first industrial revolution marked by the invention of steam engine not only brought a qualitative leap in production efficiency, but also realized the growth of economic scale and per capita output value.

According to the context of the global impact of the first industrial revolution, it is evident from Table 2.1 that the spread and application of steam engine and other new technologies have brought about significant changes in economic growth rate. Before and after the Industrial Revolution in the UK, the growth rate of gross output increased from 5.3 to 28.2% and the growth rate of per capita output increased from 2 to 12.1% in ten years, which were 5.5 and 6 times higher than the original speed respectively. Germany and Italy grew by a similar margin. After the Second World War, modern technology developed at an unprecedented speed and developed countries presented a prosperous scene. From 1961 to 1969, the United States experienced a sustained and rapid economic growth rarely seen in history, with the average annual growth rate of the manufacturing industry reaching 7.5%. Federal Germany and Japan also took advantage of the great opportunity of the post-war technological revolution to introduce a large number of advanced technology and equipment through learning, absorption and innovation, make full use of the advantage of being a late comer in order to realize the rapid economic growth and quickly leap into the ranks of developed countries.

Table 2.1 Economic growth rates of developed countries in different periods

Countries	Periods	Interval time	Growth rate per decade	
			Total output value	Per-capita output value
Britain	1700–1780	80	5.3	2.0
	1780–1881	101	28.2	12.1
	1855–1859 to 1957–1959	101	21.1	14.1
Germany	1851–1855 to 1871–1875	20	17.6	9.2
	1871–1875 to 1960–1962	88	31.1	17.9
Italy	1861–1865 to 1898–1902	37	9.7	2.7
	1898–1902 to 1960–1962	61	26.8	18.7
The United States	1893 to 1960–1962	69	42.5	17.2
Japan	1879–1881 to 1959–1961	80	42.0	26.4

Source Websites of national statistical offices

2.3.2 *Structural Effect of Technological Progress on Economic Growth*

The course of human society shows that economic growth is closely related to industrial structure. In the long historical period with the primary industry as the main economic body, the level of economic growth of human society remained at a very low level, especially from the perspective of per capita output. After the Industrial Revolution, relying on the continuous progress of technology, the rate of economic growth has been effectively improved and the total economic volume has achieved a great growth. The economic development after World War II shows that with the continuous improvement of the level of technology, it has made more contributions to human production activities. The proportion of technology intensive industries in the industrial structure has increased. People engaged in production practice by using more advanced machinery and equipment, and the proportion of labor-intensive industries and primary industries was becoming smaller and smaller. At the same time, the rise and rapid development of the tertiary industry has become a new industrial power of economic growth. It is visible that technological progress is the foundation that forms industrial labor division.

One important way for technological progress to promote economic growth is changing the input and output structure of social production resources through technological diffusion, which can also realize industrial growth through the structural effect between departments. Generally, technological progress firstly affects the departments that directly apply technological innovation and makes them form leading industries. The leading industrial sector promoted by technological progress drives the development of a series of related industries through forward and backward links so as to stimulate the sustainable growth of the whole national economy. This change in industrial relevance is mainly manifested in the form of structural changes within the original industry and various emerging industrial sectors. The continuous replacement of leading industrial sectors based on technological revolution is the deep-seated reason for the profound transformation of industrial structure. Rostow believes that: "If a society wants to maintain a high average growth rate, it must constantly carry out the anti-deceleration struggle. Modern technology attempts to resist Ricardo's law of diminishing returns. Therefore, sustainable economic growth requires the existing leading departments to continuously introduce new technology and directly or indirectly drive economic growth through the diffusion of technological innovation and its related effects in the leading departments."

Technological innovation and technological diffusion determine the change of industrial structure in economic growth, which is mainly reflected in the following aspects. As a permeable element, technological progress can expand people's ability to recognize and use natural resources and improve the effect of using original natural resources. At the same time, it will also improve the quality of labor and raise the marginal productivity of labor. The result will not only lead to differences in labor quality between industries, but also change the structure and quantity of capital, and promote the emergence and development of new products and even new industrial sectors. Therefore, technological progress will change the flow pattern of factors of production between industries and rationalize the industrial structure. From the law of economic structure change, technological innovation is first produced in the leading sectors of the economy and promotes the continuous upgrading of the entire industrial structure.

In terms of technology alone, the structural effect of technological progress on economic growth is essentially reflected by adapting or changing the economic ratio between human resources and material capital. The increase of human resource input involves two problems. One is without major changes in knowledge and skills, the increase of pure labor input can be reflected by changes in the number of workers employed in practice. Another one is human resource itself has the accumulation effect, the increase in the proportion of advanced and complex labor input is brought about by the increase in the quality, skills and knowledge of workers. The increase of material capital also involves two aspects: the increase of input under the condition of machine and technology improvement and the increase of input under the condition that technology remains unchanged. Obviously, there is a certain proportional relationship between the input of human resources and the addition of material capital. Maintaining a reasonable proportion between the input of human resources and the investment of material capital and promoting their mutual adaptation and synchronous increase can accelerate the proportional distribution of social labor in all sectors of social production, which will stimulate the coordination and optimization of economic structure, finally realize economic growth.

2.3.3 Technology, Labor Division and Scale Economies

The effect of technological progress on economic growth can be realized by promoting the development of labor division and cooperation. As far as the improvement of material capital in technological progress is concerned, the emergence and development of machines are reflected by the refinement of labor division. Without detailed labor division, it is difficult for machines to play the role. Besides, social division itself is a kind of productive forces, labor division and collaboration can produce new social productive forces. Marx pointed out that the division of handicrafts in the workshop forms the qualitative division and quantitative ratio of the social production process through the decomposition of handicraft activities, the specialization of labor tools, the formation of local workers and the combination of local workers in a general organization, thus established the certain organization of social labor and the labor productivity of the society. The deepening of labor division can also promote the development of specialization, realize increasing returns and accelerate the development of social division and the spontaneous evolution of social economic structure. The continuous division and specialization of industries is an essential part of the process by which increasing returns are realized. At the same time, the specialized labor division between regions can also promote the upgrading of industries from labor intensive to capital intensive and stimulate the technological progress of the industry. Liang and Zhan (2006) found that compared with industries with lower level of specialization, industries with higher level of regional specialization had significantly higher growth rate of technological progress and contribution rate of technology to economy.

Scale economy is the product of socialized mass production and an essential link of economic growth, which cannot be separated from the support of technology and process. The development of technology and craft determines the emergence of scale economies in different industries. The first industrial revolution, marked by the invention and application of steam engine, improved the power system of production, promoted the integration and expansion of production scale and made textile and other industries achieve rapid development. In the second industrial revolution, the invention and application of electricity provided more possibilities for mass production. After World War II, the invention and application of nuclear energy and other technologies drove the development of the energy industry. With the emergence of electronic computers, the information processing capacity of human beings increased rapidly, which promoted the development of production automation. It is the expansion of enterprise production that scale effect can be achieved. The accumulation of capital or investment played an important role in the scale economies. As Marx pointed out that in the same production conditions to invest more capital, but on one point in quality characterized by the reproduction conditions has great efficiency. Therefore, the rate of increase in production is more than the rate of increase in capital in the expansion of reproduction, that is, accumulation.

Another important reason for scale economies is that mass production can lower the cost per unit of product. Generally speaking, due to the specificity of assets, under the condition of constant fixed capital, the more the production scale expanding leads to more the output, the lower the cost included in the unit product, then the higher the use efficiency of fixed capital achieve such as machinery. Scale economy effect is a symbol of the continuous improvement of production capacity, and also an important condition for the increasingly abundant social products.

2.4 China's Economic Development Supported by Technology

At present, China is still in the stage of economic development supported by technology. Since the founding of the People's Republic of China, especially in the past 30-plus years of reform and opening up, the reform of the scientific and technological system and scientific and technological undertakings have made rapid progress and achieved great achievements unmatched in any period in history. Technology has played an important role in promoting China's economic and social development, which also follows the internal laws and external needs gradually into the right track, towards a more brilliant peak. At the same time, some specific problems arise. Therefore, this part analyzes in detail all aspects of China's technology field at the present stage and summarizes the problems, which is of great significance to form the strategy of technology promoting economic development.

2.4.1 Investment in Scientific Research

2.4.1.1 Fund Input

Since the 1990s, through the implementation of a series of science and technology plans and other policies, such as the National 973 Plan, 863 Plan, major science and technology breakthrough Plan, knowledge innovation Project and natural Science Fund funded projects, the government has continuously increased the intensity of science and technology input to drive the whole society to carry out technological innovation. It has vigorously promoted the development of science and technology and the implementation of the strategy of "rejuvenating the country through science and education". As shown in Fig. 2.1, since 2012, China's financial expenditure in the field of science and technology has shown a trend of increasing year by year, except for a decline in 2020. The proportion of fiscal expenditure in the field of science

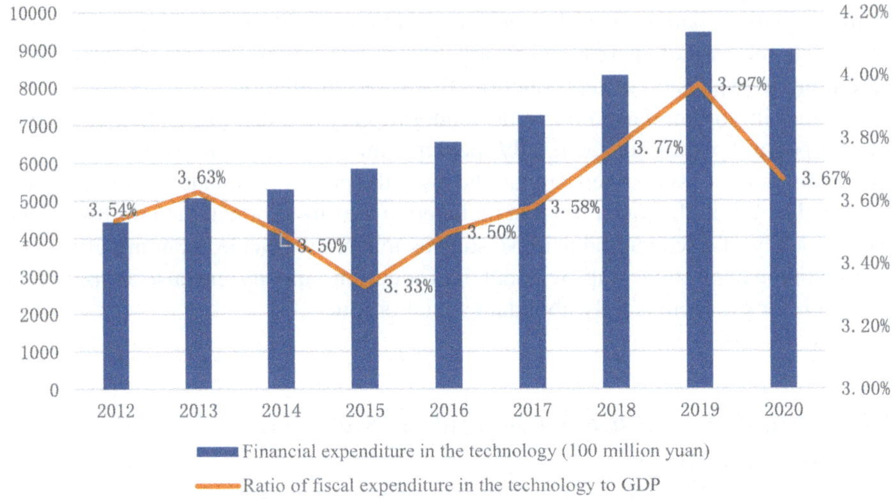

Fig. 2.1 Fiscal expenditure in China's science and technology sector from 2012 to 2020

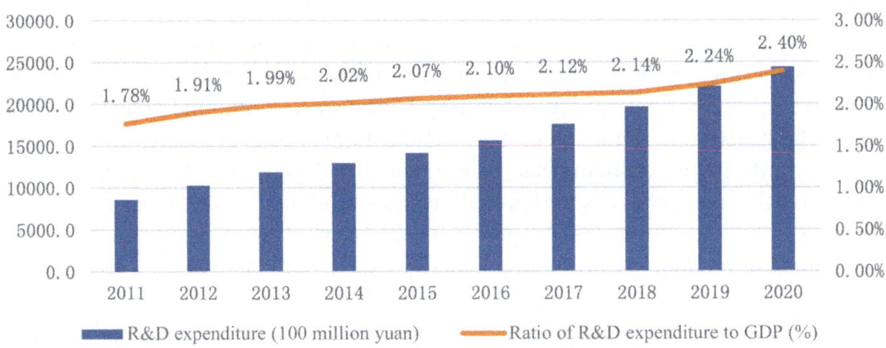

Fig. 2.2 2011–2020 R&D investment scale and intensity in China. *Source* China Science and Technology Statistical Yearbook 2011–2020

and technology in GDP remained at about 3.5% from 2012 to 2014, and reached the lowest point in 2015. Since then, the scale of fiscal expenditure in the field of science and technology has been increasing year by year. From 2011 to 2020, China's R&D investment has also shown an increasing trend year by year, and the proportion of R&D investment in GDP has also shown an increasing trend year by year, indicating that China attaches importance to R&D investment and continues to inject capital power into the science and technology field (Fig. 2.2).

Although China's financial expenditure in the field of science and technology has expanded to some extent and the investment in R&D has increased, the overall investment in science and technology economic expenditure lagged behind the demand of economic development. Since the reform and opening up, China's economic growth rate has achieved a high-speed annual growth of 8%. However, the proportion of fiscal expenditure and R&D investment in GDP was far behind the GDP growth rate and the intensity of China's science and technology investment was relatively insufficient, which did not match the situation of economic development.

In addition, compared with other developed countries, China's research expenditure was at a relatively backward level. As shown in Fig. 2.3, the proportion of scientific research investment in GDP of the United States, South Korea, Japan and Germany in 2017 was significantly higher than that of China. Among them, the proportion of scientific research investment in South Korea reached 4.15%, 1.93 times that of China. Only France and the UK match China's spending on research. Overall, there is still a gap between China and developed countries (Fig. 2.4).

The distribution of research subjects in China is also not reasonable. According to the data of 2018, in terms of foreign situation, the United States, South Korea, Japan,

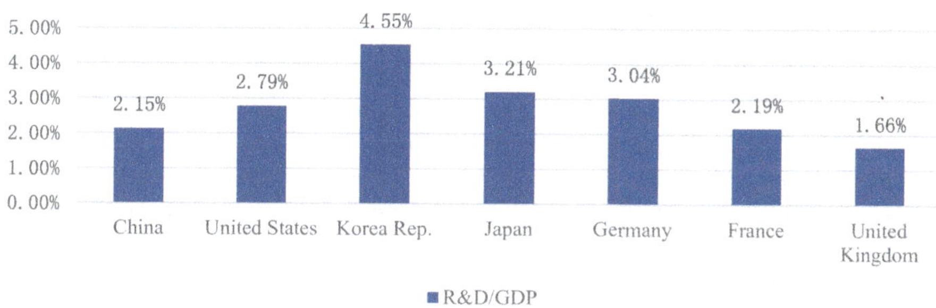

Fig. 2.3 R&D/GDP of some countries in 2017. *Source* International Statistical Yearbook

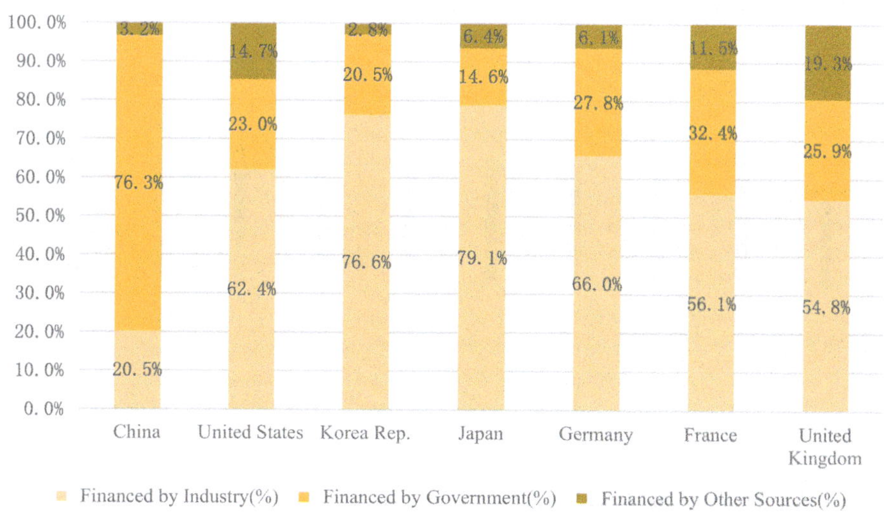

Fig. 2.4 Distribution of national science and technology funds among the three types of institutions in 2018. *Source* International Statistical Yearbook

Germany, France and the United Kingdom are mainly funded by enterprises, among which 79% of Japan's R&D funding comes from enterprises. 76.3% of China's R&D funds come from the government, which is closely related to the fact that China is a government-oriented innovation country. However, as the subject of innovation, enterprises do not have enough capital guarantee, so they do not play a corresponding role in the market and cannot meet the market demand for reasonable innovation.

2.4.1.2 Technical Personnel Structure

Talent is fundamental to scientific research. At present, China has a relatively complete scientific research team and the full-time equivalent of China's R&D personnel in 2020 is 5.235 million per year. It has sufficient personnel reserve in basic research, applied research and experimental development. As shown in Fig. 2.5, the total number of Researchers in China has been increasing year by year since 2011.

From the three aspects of basic research, applied research and experimental development, the distribution of researchers in China is unbalanced. The number of researchers involved in experimental development far exceeds that involved in basic and applied research. The unbalanced distribution structure of three types of R&D personnel is not conducive to the development of basic research. Capital and talents are the foundation of scientific research, and China is showing a trend of "attaching importance to application rather than foundation". In 2020, the number of experimental development personnel was up to 4.16 million, while the number of basic research personnel and applied research personnel was only 42.7 and 643,000 respectively, accounting for 8.2 and 12.3%. The lack of support from basic researchers will affect the efficiency of applied research and the transformation of research results, which is not conducive to the stable development of science and technology in China.

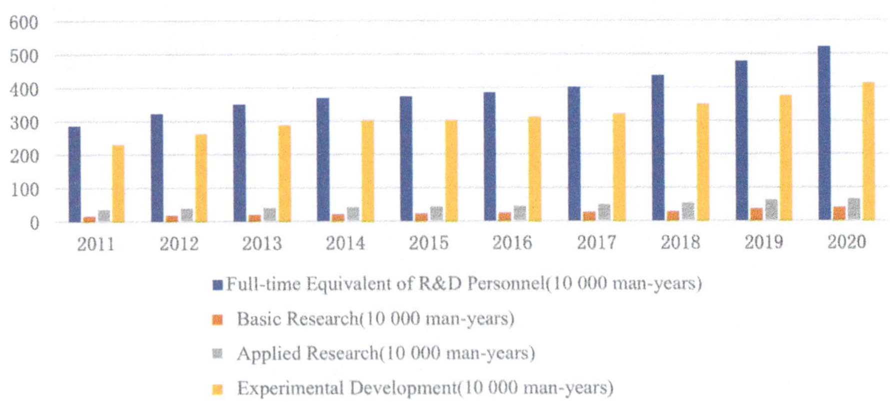

Fig. 2.5 Total number and distribution of Researchers in China from 2011 to 2020. *Source* China Science and Technology Statistical Yearbook 2011–2020

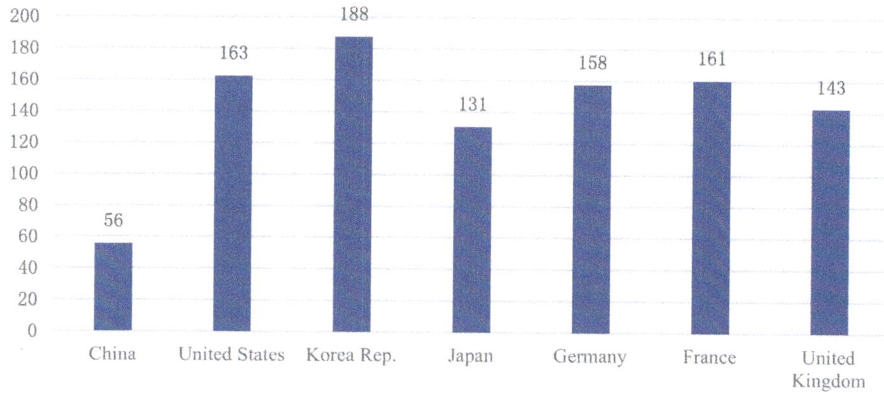

Fig. 2.6 Number of researchers per 10,000 employed persons in selected countries in 2018. *Source* International Statistical Yearbook

Compared with other countries, the number of researchers in China is still relatively small. According to the number of researchers per 10,000 employed people in 2018, South Korea has 188 researchers per 10,000 employed people, which was the highest among advanced countries, followed by the United States with 163, Germany with 158, France with 161, and the United Kingdom with 143. There are only 56 researchers per 10,000 employed people in China, a significant gap between China and developed countries. The lack of high-skilled personnel will seriously affect the pace of construction of China's manufacturing industry (Fig. 2.6).

2.4.2 *Scientific Research Output*

2.4.2.1 The Output of Papers Has Decreased and Its Quality Needs Improved

At present, internationally recognized indexes to judge the academic value of papers mainly include Science Citation Index (SCI) and engineering Index (EI). The Science Citation Index (SCI) is the most authoritative evaluation system for basic and applied research in the world. Engineering Index (EI) is a world-renowned literature retrieval system in the field of engineering and technology, which includes relevant contents in the field of engineering and technology, such as civil engineering, environment, petroleum, machinery, electronics, chemical engineering, agriculture, etc.

In recent years, the number of papers published by Chinese scientists and technicians included in these two indexes showed a decrease in volatility. As shown in Table 2.2, since 2013, the number of domestic published papers both included in SCI and EI index has declined. The number of SCI papers published by Chinese researchers accounted for 11.5% of the total number in 2013, but only 5.7% in 2018, a significant

Table 2.2 Number and proportion of Chinese papers published in SCI and EI from 2013 to 2018

Year	Taken by SCI	As % of total	Taken by EI	As % of total
2013	22,125	11.5%	54,098	35.2%
2014	24,089	10.2%	54,866	33.5%
2015	23,407	8.8%	61,873	30.3%
2016	21,789	7.5%	55,263	25.9%
2017	21,331	6.6%	47,545	22.2%
2018	21,480	5.7%	48,630	19.5%

Source International Statistical Yearbook

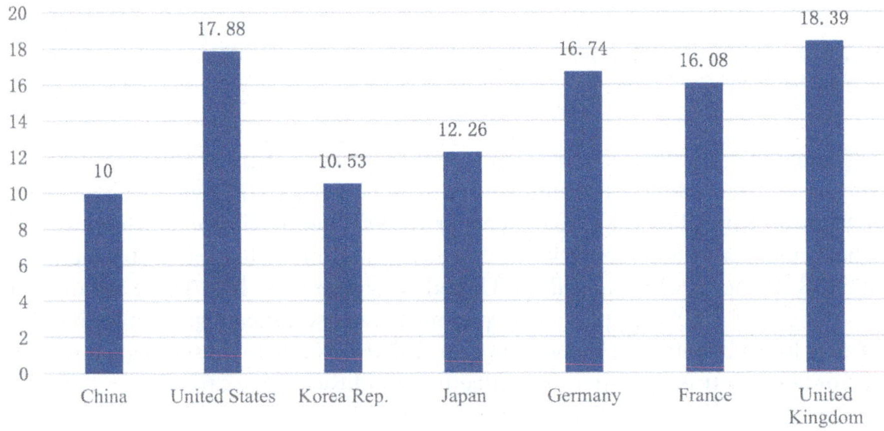

Fig. 2.7 Citation rates of some countries and some countries in 2018 (times/paper). *Source* International Statistical Yearbook

decline. The proportion of papers published on EI decreased from 35.2% in 2013 to 19.5% in 2018.

In addition to the decline in the number of research papers, the quality of papers needs to improve. As shown in Fig. 2.7, the citation rate of papers in China was only 10 times per paper in 2018, which was significantly lower than that of other developed countries. The citation rate is as high as 17.88 times per paper in the US and 18.39 times per paper in the UK. Among Western European countries, France and Germany also recorded 16.08 and 16.74, respectively. In Asia, the citation rates of Japan and South Korea were slightly higher than that of China, 12.26 times and 10.53 times, respectively. Generally speaking, the quality of Chinese papers still needs to be further improved.

2.4.2.2 Number of Patent Applications and Grants is Growing Rapidly While Poor in Innovation Efficiency

A patent is a means of protecting a technical invention. Due to the close connection between patent and invention, patent can be used to represent the output of technological invention activities to a certain extent, so patent data is used as an indicator of technological output. The patent index can reflect the invention activities of enterprises or countries, the ownership of intellectual property rights, the level of technological development and the competition between technology and economy at different macro or micro levels.

As shown in Table 2.3, the number of patent applications and grants in China increased year by year from 2016 to 2020. However, there is a large gap between the number of patent applications and patent grants, and the efficiency of innovation activities is not high. The number of patent applications reflects the active degree of technological innovation activities, while the number of patent grants reflects the efficiency and effect of technological innovation activities. From the comparison of the number of patent applications and the number of patent grants in China in recent years, the gap between them has been widening year by year, reflecting the low efficiency of scientific and technological invention activities in China.

Innovation has always been the driving force of the development of science and technology, but most Chinese scientists and technicians lack the consciousness of innovation. The total number of people engaged in research and development and the total number of enterprises engaged in research and development in China are among the top 5 in the world, but the international competitiveness of scientific and technological research and patent index is behind 20 or 30. This is mainly because China's current scientific and technological innovation system and operation mechanism is not reasonable, coupled with the government and enterprises of science and technology development strength is not strong, leading to the lack of enthusiasm for innovation of science and technology personnel and invention patent has become the "import" output way.

Table 2.3 Number of patent applications and corresponding number of invention patents in China

Year	Number of patent applications(piece)	Inventions	Number of patent grants(piece)	Inventions
2016	3,464,824	1,338,503	1,753,763	404,208
2017	3,697,845	1,381,594	1,836,434	420,144
2018	4,323,112	1,542,002	2,447,460	432,147
2019	4,380,648	1,400,661	2,591,607	452,804
2020	5,194,154	1,497,159	3,639,268	530,127

Source China Science and Technology Yearbook

Table 2.4 Status quo of high-tech products import and export in China

Year	Total value of export and import of high and new-tech products (USD 100 million)	Export	Import	Export/total value
2011	10,120	5488	4632	28.9
2012	11,080	6012	5069	29.3
2013	12,185	6603	5582	29.9
2014	12,119	6605	5514	28
2015	12,046	6553	5493	28.8
2016	11,272	6036	5236	29.9
2017	12,515	6674	5840	30.9
2018	14,185	7468	6717	31.4
2019	13,685	7307	6378	29.2
2020	14,584	7763	6821	30

Source China Science and Technology Yearbook

2.4.2.3 The Technical Content of Export Products

China's high-tech industry has a certain international scale, plays an important role in the adjustment of China's industrial structure and has become an important force supporting the sustainable and healthy development of China's manufacturing industry. As shown in Table 2.4, from 2011 to 2020, the total import and export volume of high-tech products in China has been increasing, and the import and export volume has also been increasing year by year. However, the proportion of the export of high-tech products in the total import and export has not changed significantly in the past 10 years, basically stable at about 30%, which indicates that the technical content of China's export commodities needs to be improved. We are glad that the export of technology industry reflects the international competitiveness of a country's high-tech industry. Therefore, In the future, China should pay more attention to the improvement of technological content as well as the increase of the number of high-tech industries.

References

Kuang, Guoniang, and Hanqing Tang. 2006. On Technology, Technology Diffusion and Economic. *Growth* 213 (12): 323–325.

Liang, Qi and Yijun Zhan. 2006. Regional Specialization, Technology Progress and Industrials Upgrading: Proved by Yangtze River Delta. *Economic Theory and Business Management* (1): 56–62.

Solow, Robert M. 1957. Technical Change and the Aggregate Production Function. *The Review of Economics and Statistics* 39 (3).

Chapter 3
Structure and Transformation

The first part of this chapter expounds the concepts of industrial structure and industrial transformation, explains the characteristics of industrial structure and grasps the research ideas of analyzing industrial structure and industrial transformation from an overall perspective. The second part mainly summarizes the original theories of industrial transformation and the theories of industrial structure adjustment from the perspective of macroeconomic development and endogenous economic development. The third part explores the law of industrial structure change and economic development from the perspective of historical development, and obtains the direction and path of future industrial development from the perspective of industrial transformation. The fourth part analyzes the current situation of China's industrial structure from the perspective of the distribution of three major industries, the change of employment structure and the development of traditional and emerging industries.

3.1 Related Concepts

3.1.1 Industrial Structure

Industrial structure refers to the proportion of production factors among various industrial sectors and the interdependence and mutual restriction between them. In other words, it is the distribution of a country's labor force, wealth and resources among various industrial sectors of the national economy and the way they restrict each other. In modern society, industry, agriculture, construction, transportation, post and telecommunications and service industry are regarded as basic industrial sectors. Industrial structure plays a leading role in the whole national economic structure.

For the industrial structure, people often use the proportion of stock scale. Different indicators are used to measure the stock scale, resulting in different characteristics of industrial structure. Some people use the distribution proportion of labor force, some use the material form of labor accumulation in the past, the proportion of

Y. Jiang and J. Gu, *Technology and Industrial Transformation of China*,
https://doi.org/10.1007/978-981-19-7458-8_3

fixed assets, and still some use the combination of labor force and production means, the proportion of output value (income). The structural parameters represented by the proportion of stock scale provide a reference for understanding the trend and direction of production evolution and clarify the action policy of planning industrial structure, essential rationality. For example, the changes of the three industries with the improvement of economic level have a generally consistent trend in different countries, pointing out the basic direction for production planning.

However, the industrial structure of different countries has various experiences in the process of developing in this basic direction, which shows that different countries have taken different behavioral steps, that is procedural rationality in the process of achieving essential rationality. From essential rationality to procedural rationality is not only the process from the description of ideal state to real economic life, but also the process from industrial structure theory to industrial policy (decision-making).

3.1.1.1 Characteristics of Industrial Structure

Structure refers to the various elements that make up something and the way they relate to each other. Both natural and social phenomena are composed of various elements in a certain way. With limited conditions, different quantities of the same element, or different combinations of the same quantity, will have different structures, reflecting variant properties of things. If a country or region adopts a structure dominated by agricultural economy, generally speaking, the economy of the country or region is underdeveloped and the level of productivity is not high. The economic structure dominated by industry has a higher degree of economic development than that dominated by agriculture.

Every system has a structure. A system is an organic whole with specific functions composed of several elements that interact and depend on each other. In the complex social and economic structure, many factors are not simply added, but are interdependent and restricted according to certain laws. Industrial structure is a subsystem of economic structure and an organic whole. Its performance is that each element of the industrial structure has a certain level, correlation and relativity, which is not a simple accumulation.

First, the industrial structure is hierarchical. Everything has a quantitative change to qualitative change, from low to high development process, has its progressive and sequential. In different stages of development, there are different sequences and levels. In the areas with low productivity and underdeveloped commodity economy, the rural industrial structure is often a single agricultural structure. In areas where the commodity economy has developed greatly, the rural industrial structure often forms a combination of industry and agriculture. With high productivity and developed commodity economy, a multi-sectoral comprehensive industrial structure is often formed. The development process from unitary industrial structure to integrated industrial structure reflects the three stages of the development of China's rural industrial structure. The formation of industrial structure is restricted by rural economic structure, national economic structure, industrial policy, market consumption structure, natural geographical environment and other factors, which form a joint force and restrict the rural industrial structure from various aspects.

The hierarchy of industrial structure is manifested in many aspects and affects the level of macroscopic decision making. The first level is the primary, secondary and tertiary industrial structure. The second level is the internal structure of each industry, that is, the agricultural, forestry, animal husbandry and fishery structure of the primary industry, the industry and construction structure of the secondary industry, the circulation structure, service industry structure and management structure of the tertiary industry. The third level is detailed to the sectoral structure of various industries, such as planting structure, forestry structure, animal husbandry structure, light and heavy industry structure in rural industry, commerce and transportation structure in circulation. A rational industrial structure needs to be layered. As the foundation of national economy, agriculture is also the foundation of establishing reasonable industrial structure. Therefore, in adjusting the agricultural industrial structure, we should adjust the agricultural structure in the first level. In the process of adjusting the secondary industry, we should pay attention to adjust the extensive industrial development model, not only to obtain the best economic benefits, but also to maintain the unity of social, ecological and economic benefits. In tertiary industrial structure, we should first start with adjusting the circulation, production and living service departments. In view of the immature development of China's tertiary industry, we need to adjust step by step, integrate the development of industries at other levels and pay attention to the relationship between various levels.

Second, industrial structure has correlation. The relevance of industrial structure is not only reflected in the linkage and feedback between the elements of industrial structure, but also between elements and structures. Only by connecting the elements and links constituting the structure for comprehensive analysis, and comprehensively investigating the correlation between elements and elements and between elements and structure from both qualitative and quantitative aspects, can we grasp the rationality of industrial structure as a whole. What plays a key role in the industrial structure is often determined by the weakest factor in the structure, which is also known as the law of limiting factors. In production, if there are sufficient raw materials and lack of corresponding mechanical equipment, the raw materials cannot be processed into new products. At this time, the production and quantity of machinery determine the overall function of the structure. Therefore, the fundamental strategy of adjusting the industrial structure is to coordinate the overall elements and can't make the absolute situation between the elements.

Third, industrial structure is relative. The elements in the industrial structure are not isolated and the establishment of a reasonable industrial structure is a dynamic rather than a static development process. The industrial structure is not a rigid model nor does it keep a fixed proportion for a long time. It needs to change in response to the needs of the society, and will often change with the improvement of production conditions and living standards, which is a flexible and reasonable structure. The rational industrial structure will not become a fixed template due to the difference of time, place and conditions. Comparing the development modes of Jiangsu province and Guizhou province, both are developed from the past, but there are significant differences in the focus of regional industrial structure. Obviously, the elastic industrial structure does not deny its relative stability. The changes of various elements forming the industrial structure in a certain period of time begin with quantitative changes, that is, no substantial structural changes have taken place. At this time, the industrial structure has relative stability. Therefore, we should not only see the relative rationalization of industrial structure, but also the relative stability of industrial structure, which is the basic starting point of formulating industrial structure.

3.1.2 Industrial Transformation

High-quality development should achieve a relatively complete industrial system, networked and intelligent production organization, strong innovation, demand capture, brand influence, core competitiveness, high-quality products and services. In 2019, General Secretary Xi Jinping stressed that "manufacturing is an important carrier of the development of the real economy, and we must pay attention to the development of China's manufacturing industry and the real economy." In fact, industrial transformation has become the key to promoting supply side structural reform and building a modern economic system.

The meaning of industrial transformation has also experienced different levels of evolution. The first level of industrial transformation is discussed from the change of the internal structure of a country's national economy. William Petty and Fisher explained industrial upgrading from the perspective of three industrial structure changes earlier. Thus, the three industrial divisions of primary industry, secondary industry and tertiary industry were gradually opened. Clark further verified the evolution law of industrial structure on the basis of Fisher's three-industry division method and the research on the relationship between income and labor mobility, namely the "Petty-Clark Theorem". He believed that with the increasing of per capita national income, the proportion of the secondary industry would gradually surpass the primary industry and become the mainstream industry. After a period of economic development, the tertiary industry occupied the main position of economic development, while the proportion of the primary and secondary industries decreased relatively. Thus, the first-level understanding of industrial transformation was initiated, that is, the higher the proportion of the tertiary industry, especially the proportion of service industry in GDP, the more advanced the economic development level and the better the industrial structure.

The second level of industrial transformation is interpreted from the perspective of openness. This was an evolutionary process that appeared simultaneously with the process of industrialization and the opening up of a country and developed with the acceleration of China's opening up. Akamatsu proposed Flying Geese Pattern, arguing that the post-production countries have to go through three development stages of "import, domestic production and export" in order to effectively achieve the dual goals of economic development and industrial transformation and upgrading.

The path to realize industrialization of the latecomer countries is through the industrialization of the first country and then the realization of heavy industrialization to high processing degree. At the same time, a country's production and consumption cycle will also go through the process of light industrial products and then heavy industrial products. Yamazawa extended the theory, arguing that the backward countries need to introduce advanced technologies and products from developed countries to achieve economic take-off, and described the three stages of Akamatsu as five stages: "introduction, import substitution, export growth, maturity and reverse import", which shows a country's industrial transformation and upgrading needs to be realized through opening up.

Kojima (1986) considered the importance of FDI in the process of industrial transformation. He argued that countries such as Japan need to upgrade its industries by bringing in foreign investment or gradually moving weak domestic industries abroad. Since then, the space for industrial transformation has been further extended from domestic production, consumption and export to overseas production. From the perspective of industrial value chain, Gereffi (1989) believed that industrial upgrading is actually a process of evolution from low technology and low added value to high technology and high added value. Porter (2003) regarded industrial upgrading as a process in which a country's comparative advantage of factors promoted the development of labor, land and other resource-endowed industries to capital and technology-intensive industries. Nicholas (1985) put forward the view that industrial policy linkage would indirectly promote regional industrial transformation. Mathews emphasized the influence of policy and institutional environment on industrial upgrading. Taking Asian economy as an example, he believed that institutional policy factors would promote the global transfer of new technologies.

The above research is of great value to understand China's industrial transformation. For example, the understanding of the status of the three industries, since the outbreak of the financial crisis in 2008, the increase of the proportion of the tertiary industry in GDP has been regarded as a symbol of the advanced industrial structure of a country. After the financial crisis, the secondary industry with manufacturing as the core has received new attention. The 14th Five-Year Plan of many provinces in China explicitly stipulated that manufacturing should keep an adequate proportion in GDP. Another example is how to promote China's industrial transformation and upgrading in the context of an open economy in which domestic and international double cycles promote each other.

3.2 Existing Theories of Industrial Structure Adjustment

The research on the evolution of industrial structure has always been the key content of industrial economics, mainly including the internal relationship between industrial structure evolution and economic growth from the perspective of macro structure, the evolution law of the three industrial structures, the internal relationship between industrial structure evolution and economic growth from the perspective of endogenous economic growth and the process analysis of industrial structure evolution perspective analysis of the process.

3.2.1 The Macro Perspective

The process of industrial structure change is the basic characteristic of economic development. The research on the evolution of industrial structure from the perspective of macro structure mainly focuses on the evolution of three industries and the internal structure of industry and the inherent regularity of economic development stage. These studies began with William Petty's research on the relationship between industrial structure difference and national income difference. With the deepening of research, scholars generally agree that the process of industrial structure change is the basic feature of economic development and an important variable to explain the overall productivity growth and development direction (Kuznets 1967; Pasinetti 1984). Some research works involved are as follows: first, exploring the influence mechanism of three industrial structures on national income and economic growth, such as demand income elasticity and structural dividend hypothesis (Timmer and Szirmai 2000; Gan and Zheng 2009), different tendencies of entrepreneurial activities in different industries and knowledge spillover of producers or users (Peneder 2010); second, the law of the change of industry internal structure, main point of the above mentioned and 40s of the twentieth century Kaname's theory of "flying geese pattern" and the fifties and sixties of the Leontief input-output analysis, Rostow's theory of leading industry diffusion effect and the theory of economic growth stage, Chenery's industrialization stage theory and Housman's industrialization experience law. These views explain that with the evolution of economic development stage, the internal structure of industry changes in the following order, light industrialization, heavy industrialization, high processing and knowledge and technology intensification of industrial structure.

As far as domestic research is concerned, it is indisputable that industrial structure change has a positive effect on economic growth and development mode, but there are also differences in views on some issues, mainly reflected in the phased position of industrial structure change in the time series of economic growth, the difference in the impact of industrial sectors on economic performance and the impact of industrial structure characteristics on economic growth.

As to the role of industrial structure changes and institutional reforms, Liu (2003) study shows that the upgrading of industrial structure caused by dual economic transformation is higher than the economic institutional change in terms of both the improvement of growth quality and the marginal contribution to economic growth. Liu and Zhang (2008) study shows that in the 30 years since reform and opening up, although the contribution of industrial structure change of China's economic growth was significant, but with the improvement of marketization degree, the industrial structure change contribution to economic growth trends, to decrease gradually give way to the technological progress and institutional change. Similarly, Mao and Pan (2012) pointed out that the effect of industrial structure on the improvement of urban labor productivity in China shows an inverted U-shaped change. Ruogu et al. believed that industrial structure and system not only have a direct impact on economic scale, but also affect the output efficiency of production factors through their function of resource allocation, thus having an indirect impact on economic growth. These research viewpoints show that both industrial structure change and institution affect the change of overall productivity, and their influence degree is characterized by obvious stage evolution. Moreover, both industrial structure change and institution have different short-term and long-term effects on economic growth and factor allocation function.

The impact of industrial sectors on economic performance is different. According to Liu and Li (2002), the most effective industry to drive economic growth in China is almost tertiary industry except the construction industry. Scholars also have different views on the direction of industrial structure adjustment. One holds that high-tech industry should be the driving force, modern service industry and modern manufacturing industry should be the two wheels of development to lead the overall upgrading of industrial structure (He and Yao 2008). The other believes that China should promote industrialization led urbanization, while avoiding excessive urbanization caused by premature servitization of industrial structure (Qiao et al. 2012). At the regional level, the study of Zhou and Wu (2008) shows that the gap between the proportion of industry, especially manufacturing, in GDP and industrial competitiveness is the main reason for regional employment differences.

For the influence of industrial structure characteristic to economic growth, there are few research works on acting in China. The main research points include industrial linkage mechanism, technology selection effect (Huang and Li 2009), industrial structure rationalization, while the role of industrial structure upgrading is uncertain (Fu 2010).

3.2.2 The Perspective of Endogenous Economic Growth

Endogenous economic growth theory emphasizes the importance of division of labor, knowledge spillover, technological progress, increasing returns and so on to long-term economic growth. Becker and Murphy (1992) introduced coordination cost into the long-term economic growth model and believed that knowledge accumulation and coordination costs jointly affect the degree of division of labor and thus affect economic growth. Nelson and Kamal (2002) emphasized the importance of the co-evolution of technology, industrial structure and national system for economic growth. With the guidance of these new economic growth theories, scholars from industrial economics, regional economics and economic geography began to discuss the relationship between industrial structure change and endogenous economic growth. Frenken and Boschma (2007) regarded local economic development as a process of sector evolution of product innovation and continuous transformation of industrial structure of economic activities. The scope of product innovation depends on the diversity of existing enterprises and cities. The income generated by the diversity of these two levels is the core mechanism of the industrial sector evolution driven by product innovation.

Knowledge spillovers and externalities are important sources of enterprise innovation. In a specific space, the research on knowledge spillovers and externalities is carried out in two contexts, where one is to investigate the knowledge spillovers from organizational R&D activities, focusing on the dynamic externalities between the geographical proximity of universities and business research activities (Florida 2002; Greunz 2004). Another is to investigate the internal relationship between externalities and industrial activity components (Glaeser 2003), that is, to discuss what kind of industrial structure clustering in space is most conducive to knowledge spillover and economic growth from the perspective of industrial activity components of spatial agglomeration. There are two views on this:

The first is a specialized and diversified industrial structure. One believes that specialized industrial structure contributes to knowledge spillover, which is MAR externality. Others believe that diversified industrial structure contributes to knowledge spillover, which is JACOBs externality. These two views form the dispute of specialization and diversification of regional industrial structure. Some scholars believe that diversified industrial agglomeration is conducive to economic growth, enterprise birth and innovation (Rosenthal and Strange 2001; Li and Song 2008). Other scholars believe that specialized industrial structure is conducive to the improvement of regional labor productivity (Capello 2008; Fan and Shi 2009). Peng and Jiang showed that MAR externalities and JACOBs externalities had significantly positive impacts on regional industrial innovation in China, but the impact degrees were different: industrial complementarity in JACOBs externalities had the greatest promoting effect on innovation.

The second is related diversity of industrial structure. Henderson (1997), Frenken et al. (2007) and other evolutionary economic geographers developed the theoretical view of "Related Variety Industrial Structure". This provides a developmental and integrative perspective for the debate on "diversity" and "specialization" industrial structure. According to this view, in terms of industrial structure, it is necessary to maintain a reasonable balance between diversity and professionalism in order to facilitate the long-term development of local economy. Related diversified industrial structure can promote innovation and economic growth, while unrelated diversified industrial structure cannot learn from each other due to the lack of common knowledge base due to the large cognitive gap between enterprises, which cannot promote economic development, but can restrain unemployment or stabilize employment. Hu et al. (2011) pointed out in their empirical study on China that simply emphasizing specialization or diversity is not conducive to enhancing regional competitiveness. Sun (2020) conducted an empirical study on the level of Cities in India and found that relevant diversity has a significant role in promoting regional economic growth and economic stability.

3.2.3 The Theoretical Viewpoint of Industrial Structure Adjustment

Based on the theory of classical and neoclassical equilibrium distribution, this part analyzes the internal relationship between industrial structure change and economic growth from the perspective of macro structure through comparative static analysis. First, it regards economic growth as a quantitative difference, ignoring the heterogeneity of firms and industries, inter-industry interdependence and resource transfer between different industries. Mackay (1969) pointed out that when analyzing the impact of industrial structure on regional economic growth, the interdependence between industries, especially technological connection, should be considered. Second, the difference of national income level and the key reasons for different stages of economic development are attributed to the different proportion of industrial structure. Therefore, how to change from low productivity sector to high productivity sector has become the main research topic. It is often assumed that different industries have different productivity, without taking into account the internal factors (technology) causing the change of industrial structure. This is an inevitable defect of the analysis based on the macroeconomic perspective.

The study of industrial structure change from endogenous economic growth provides a very enlightening theoretical perspective for the study of industrial structure adjustment, which is based on the knowledge spillover effect generated by industry in spatial agglomeration. To analyze the impact of industrial structure characteristics such as industrial specialization and diversification on the development of regional economy, the view that "relevant diversified industrial structure is conducive to the innovative development of regional economy" has been empirically supported by scholars at home and abroad. However, from the perspective of endogenous economy, there are still two issues that need to be studied. First one is, for a region, what is the micro mechanism of the formation of relevant diversified industrial structure. Second one is the impact of the organization mode of industrial activities in the region on the dynamic change of industrial structure.

The existing research broadens people's theoretical understanding of industrial structure adjustment, but also has some shortcomings. It is not difficult to find that theoretical research increasingly attaches importance to the internal relationship between industrial structure and enterprise innovation. At present, the academic circle basically agrees that industrial structure adjustment is the endogenous accumulation change process for enterprises to achieve innovation goals. On the basis of this understanding, the existing research has shifted to the enterprise level, but there is still a lack of analysis on the internal logical relationship between enterprise innovation and industrial structure adjustment.

3.3 Industrial Structure Change in Economic Development

After the end of World War II, a large number of newly independent nation-states emerged on the world stage. The economic development of these developing countries has also become an important research field in economics. Among many fields involved in the theory of economic development, industrial development and economic structure change promoted by industrialization is undoubtedly one of the core topics. Therefore, a large number of economists have conducted in-depth research on this theme and achieved fruitful results.

3.3.1 The Inevitable Path for Developing Countries

The industrialization of European and American countries left a deep impression on the latecomer countries. In lots of developing countries, development means industrialization and modern industry is regarded as the symbol of developed economy. Thus, in the 1950s and 1960s, industrialization was widely adopted by developing countries as the main goal of economic development. In the early literature of development economics, industrialization was an important issue that scholars paid great attention to. Since the 1940s, economists have been discussing the industrialization of developing countries. The influential literature includes Rosenstein-Rodan's *Problem of Industrialization in Eastern and South-Eastern Europe*, Staley's *World Economy in transition* and Mandelbaum's *The industrialization of backward areas.* These literatures all believe that industrialization can promote the transformation of economic structure and economic growth in less developed countries.

The so-called industrialization is the process in which the modern industrial sector gradually replaces the traditional agricultural sector as the leading industry in the economy and promotes economic development. The process of industrialization is often accompanied by significant changes in the economic structure, which is not only reflected in the rising proportion of industrial sectors in the industrial structure, but also in the proportion of employment absorbed by industrial sectors in the overall employment structure. In the famous Dictionary of New Palgrave Economics, the word "Industrialization" has been described as "the basic characteristics of ethylene in a specific Industrialization process are as follows: First, manufacturing activity and secondary industries have generally increased as a share of national (or regional) income, except perhaps for disruptions caused by the economic cycle. Second, the proportion of the employee in manufacturing and secondary industries also tends to increase. The increase in both rates was accompanied by an increase in the per capita income of the population as a whole, except for temporary interruptions." To a large extent, this expression reflects the consensus of the western mainstream economic circles on the basic characteristics of "industrialization".

Development economists generally believe that industrialization is a necessary stage for developing countries to emerge from economic underdevelopment.

Economic development means the fundamental change of economic structure. Therefore, the law of economic structure evolution in the process of industrialization became the focus of research at that time. From the perspective of economic history, in the process of an economy from an underdeveloped state to a rapid development state, accompanied by the development of industrialization, the economic structure, especially the industrial structure, also underwent great changes. Lewis revealed the evolution of industrial structure in developing countries at the early stage of economic development by constructing a dual economic structure model. He divided the favored industrial sectors of developing countries into traditional agricultural sector and modern industrial sector. Due to the difference in labor productivity between the two sectors, labor would flow from the inefficient agricultural sector to the efficient industrial sector, and the continuation of this process led to the size of the industrial sector surpassing that of the agricultural sector. Kuznets (1975) used statistical methods to conduct cross-sectional analysis and historical analysis of 57 countries, and found the evolution law that the three industries successively occupied the dominant position in the industrial structure during the evolution process. In the initial state of economic development, agriculture played a leading role in the national economy and agricultural products occupied a large proportion in people's demand structure. As people's income level increasing due to economic growth, the demand for industrial products would rise constantly, thus encouraging the expansion of the industrial sector to continuously meet people's material needs. When people's income level increased to a certain extent, there would be more and more demand for non-material services, so the service industry developed rapidly.

In the process of industrialization, not only does the industrial structure have the rule that the proportion of the three industries increases once, but also the internal structure of the industrial sector will change. Hoffmann found that in the late industrialization, there was a trend of upgrading from labor-intensive consumer goods sector to capital and technology-intensive capital goods sector. With the improvement of income level brought by economic growth, people's demand structure changed, which would inevitably stimulate the adjustment of production structure and then promoted the change of industrial structure. At the beginning of the industrialization, in order to better meet people's basic living needs, food processing, light textile and clothing and other consumer goods sector developed faster. Because these sectors belong to labor-intensive industries and have low requirements for capital, technology and other production factors, they quickly matched the resource endowment of developing countries with scarce capital and abundant labor force, which would develop rapidly. With the deepening of the industrialization process, the capital accumulation of developing countries gradually increased and the capital-labor ratio also began to change. The requirements of expanding production scale, improving production efficiency and improving technological level constitute the internal motive force for the development and expansion of capital and technology-intensive capital goods sector. The industrial sector of developing countries is upgrading along the direction of resource-intensive, labor-intensive, capital and technology-intensive industries and the industrial structure is gradually evolving from low to high planning. However, some scholars believe that "Hoffman's Law" is not universally applicable.

Kuznets pointed out that it was impossible to draw a general conclusion from the development experience of the United States, so the formula and data of "Hoffman's Law" needed to be significantly revised.

Through reviewing the development history of the world economy, Rostow (1960) summarized the economic development process of an economy into six stages from the perspective of economic history, traditional society, the stage of creating the premise for take-off, the stage of take-off, the stage of advancing to maturity, the stage of large-scale high consumption and the stage of pursuing the quality of life. Among them, the take-off stage corresponds to the start of industrialization. In order for an economy to start industrialization and take off, some basic conditions must be met. First, capital accumulation must reach a certain level. Industrialization is inseparable from a large amount of material capital input and only when the investment rate rises from 5 to 10% or even higher can the economic growth in the take-off stage be guaranteed. Second stage focuses on developing the leading sectors of the national economy. In the initial stage of industrialization, it is unrealistic to realize the parallel development of industrial sectors. One or more important manufacturing sectors should be allowed to develop. Only by guiding the limited economic resources to the leading industry with strong correlation effect, can the industrial sector develop as a whole and promote the sustainable economic growth through the development of leading industry and its drive to the related upstream and downstream industries. Third stage promotes overall reform and creates favorable institutional and social conditions for the smooth progress of industrialization. Only by creating a political system, economic system and social structure that is conducive to the development of industrialization, can we guarantee the development of modern industrial sectors and the smooth realization of economic take-off, so we should promote comprehensive reform.

3.3.2 Protecting Infant Industries and Promoting Industrial Growth

The research of industrial structure changes in the process of industrialization pointed out the direction for developing countries to promote industrial upgrading. However, the practical problem facing developing countries at that time was that the industrial base was extremely weak and the industrial system had not been completely established in the early stage of economic development. Export earnings mainly depended on primary products and they had to import a large number of industrial products from abroad in order to meet the needs of the domestic market. Therefore, how to promote the rapid growth of their own industrial sectors has become a key problem that developing countries urgently need to solve. In the 1950s, represented by Raúl Prebisch, an economist of Latin America, due to historical reasons and the difference of economic development level, world economy formed a "core-periphery" inequality of relations between developed countries and developing countries. In the "core-periphery" trading system, with the technological progress of the industrial sector and the expansion of commodity export scale, developing countries had a long-term trend of deteriorating terms of trade. If we want to break this pattern, we should implement the economic development strategy of import substitution. Through the development of the industrial sector, replacing imported industrial products with domestic industrial products. With the continuous upgrading of domestic consumer demand, promoting the expansion of the domestic industrial sector from the general consumer goods production sector to the durable consumer goods sector, and then to the capital goods production sector, so as to achieve sustained economic growth in the process of deepening industrialization.

In order to promote import substitution strategy, it is necessary to protect infant industries. For a long time after the World War II, the theory of infant industry protection was widely accepted by developing countries. From the perspective of theoretical origin, Hamilton, the former secretary of the Treasury of the United States, was the first person who put forward that limited protection should be given to infant industries. List, the representative of The German historical school, explained this theory systematically for the first time in his famous work *the National System of Political Economy*. At that time, Germany's productivity was far behind that of early industrialized countries such as Britain. Germany, as a latecomer, was able to obtain the industrial products it needed through free trade, but List pointed out sharply that "the productivity of wealth is infinitely more important than wealth itself." If the principle of free trade was fully believed, the weak German industrial sector would have a disadvantage in the competition with the powerful British industrial sector, which was very unfavorable for the development of German industrial sector. Therefore, List believes that only the implementation of tariff protection policy can Germany get the opportunity to catch up. He also believes that tariff protection alone is not enough for the industrial sector to catch up. It is necessary to actively introduce foreign advanced technology, establish a patent system to encourage technological innovation and improve the education system to train more high-quality labor forces. List acknowledged that tariff protection harmed consumer welfare in the short term, but it compensated for it by increasing productivity. With the guidance of List theory, Germany successfully achieved industrial catch-up at the end of the nineteenth century and still maintains the status of the world's industrial power.

In the more than 100 years after List, many economists continued to perfect the theory of infant industry protection. The basic idea of this theory is that a latecomer country, in its efforts to promote its own industrialization, should provide limited protection through high tariffs for its fledgling industries with potential advantages. Because these naive industries were at a temporary disadvantage in terms of production cost compared with similar foreign enterprises, if the latecomer countries implemented the free trade policy, these industries with potential cost advantages in the future would be impacted by foreign products. Therefore, governments of latecomer countries should protect infant industries through protective trade policies. The high tariff made the price of foreign products remain high so that the cost advantage over domestic enterprises couldn't be reflected and they avoided the huge market impact. During the protection period of infant industry, with the increase of domestic demand for related products, domestic enterprises continuously expanded the production scale. With the expansion of production scale, domestic enterprises gained more production experience, the labor skills of workers were more and more skilled and the product quality was more and more stable, making the production cost of domestic enterprises reached or even lower than that of foreign competitors. When domestic infant industry had international competitiveness, infant industry protection period was over. Although the protection of infant industries through trade policies distorted market prices, affected resource allocation and led to loss of efficiency, once the immature industries grew and became an important force for the country to participate in international market competition, their excess profits would fully compensate for the welfare loss caused by protective policies.

In the 1970s and 1980s, some scholars organically integrated industrial organization theory and international trade theory to form a new trade policy and achieved great development, represented by Dixit (1976) and Krugman (1986). This theory demonstrated the necessity of trade protection for some industries with increasing scale from a new Angle. Different from the traditional trade theory based on constant returns to scale and perfect competition, the basic premise of strategic trade theory was the assumption of imperfect competition and increasing returns to scale. Strategic trade theory used mathematical model to make strict reasoning and argumentation, which proved the non-optimality of free trade policy and the rationality of government intervention in some cases. This theory stated that certain industrial sectors were characterized by dynamic economies of scale, in which production costs decreased with the increase of production scale. Moreover, because these industries were capital-intensive or technology-intensive, the formation and growth of these industries were conducive to the diffusion of technology among domestic industrial sectors. Therefore, state support for these industrial sectors was beneficial to economic growth. Strategic trade theory holds are as follows. In the case of competition from foreign companies, the government protects domestic enterprises through tariffs, subsidies and other means to occupy most share of domestic market and to be able to constantly expand the scale of production. In this process, the marginal cost of production along the "learning curve" continued to decline, gradually formed economies of scale and low-cost competitiveness and finally enable domestic low-cost goods to compete with multinational companies in the international market. In all, if a country wants to be at the forefront of industrial technology development, it must support those industrial sectors with obvious economies of scale and externalities.

3.3.3 Technological Development and Industrial Transformation

While developing countries committed to industrialization, the economic restructuring of developed countries represented by the United States appeared a new trend. The proportion of manufacturing industry in the economy declined, but the production efficiency and competitiveness of manufacturing industry increased. The industrial chain of high value-added links became the focus of the manufacturing industry development, and economic development mainly relied on the scale and speed of service industry development. Among the service industries, producer service industry is developing the fastest, of which the value-added and employment have been rising continuously. With the help of information technology, the service industry has overcome the so-called "Baumol's cost disease" of low labor productivity, and labor productivity has driven the improvement of total factor productivity of the entire industrial sector. The service sector in the US and other major developed countries accounted for more than 70% of GDP, and the service sector has become a major driving force for economic development. Developed countries were transforming from an industrial economy to a service economy.

Based on the experience of developed countries, economic development does not simply mean the industrialization of agricultural countries and the transformation of industrial structure is not only reflected in the increase of the manufacturing proportion in GDP. In recent years, emerging economies have also seen a decline in the share of secondary industry and a rise in the share of tertiary industry. Does this mean emerging economies are about to enter a new phase of service-led development? The importance of manufacturing to an economy would not gradually disappear with the completion of industrialization. After the financial crisis, developed countries rethought the importance of manufacturing industry and launched the policy of "Re-industrialization", which showed that even in the post industrialized era, the sustainable economic growth of developed countries was inseparable from the support of manufacturing industry. Hausmann showed that with the continuous development of manufacturing industry, people's production skills and technological accumulation would continue to increase and the ability to create knowledge continued to improve, which made the industrial products produced by more developed manufacturing industry more complex. Thus, a country's long-term growth prospects can be predicted from product complexity. The huge income gap between different countries is closely related to their manufacturing industry. To sum up, the development of manufacturing industry can not only create economic benefits, but also play an unmentionable role in national economic growth. From the perspective of the source of technological innovation, manufacturing industry has always been the most active sector of technological innovation. Manufacturing sector is not only the main promoter of technological innovation and technological progress, but also the important user and disseminator of new technology.

Competition is the essential requirement of market economy and the core process of market mechanism. Market competition can help stimulate technological innovation, improve corporate efficiency, drive industrial development and ultimately gain consumer welfare. The history of global industrial development since the first industrial revolution showed that innovation and competition were the main driving forces of industrial evolution. In the era of faster technological innovation and higher cost, it is far from enough to rely on enterprises to promote a country's technological innovation. Government can play an important role in promoting technological progress. Emerging industries characterized by strategic, uncertainty, positive externalities and complexity, are prone to market failure, which provides a large space for government intervention (Li and Lv 2010). In the investigation of the dynamic process and mechanism of technological change in the industrialization in East Asian countries and regions, scholars found that the active and effective government intervention determined the rate and direction of technological progress in later countries. Through his research on Japan's technological changes, Freeman found that Japan could achieve comprehensive economic catch-up and technological leapfrog from the ruins of the World War II because the development strategy of establishing the nation through technology and the technological progress policy promoted the technological innovation of enterprises significantly. Thanks to the encouragement of the Japanese government and the support of industrial policies, Japan gradually formed a national technological innovation system integrating industry, government and education, which effectively supported Japan's technological progress and economic growth.

After economic development entered a new stage dominated by technological innovation, the role of government intervention in industrial development also underwent new changes. Improper government intervention would neither promote technological progress and industrial transformation nor stimulate industrial growth. For a long time, governments in East Asia promoted rapid growth of industrial sectors by selecting leading industries and supporting them vigorously. This kind of government-oriented development strategy and industrial policy may be more effective in the early stage of economic development because of the industrial development experience in developed countries. However, with the upgrading of domestic industry and the increasing complexity of economic structure, the limitation of government cognitive ability implies the possibility of forecasting errors. If the government makes a big mistake in judging the next stage of strategic industries, the policy of selecting and supporting leading industries will inevitably lead to the waste of resources.

3.4 Development of China's Industrial Structure

3.4.1 The Formation of a Service-Oriented Economy

Since 2000, China's economy has shown a trend of rapid growth. GDP grew 8.5% in 2000 and 10% in 2003, a pace that continued until 2007, when it reached an unprecedented 14.2%. In 2008, the economy slowed down due to the global financial crisis. Since 2010, China's national economy has bid farewell to the era of high-speed growth and presented a "new normal" of medium–high growth. In 2014, China's GDP grew 7.4% from the previous year, decreasing 0.4% points. Since then, the economic growth rate has been maintained around 6.0–7.0% from 2015 to 2019. In 2020, due to the impact of the global COVID-19 pandemic, the economic growth rate recorded the lowest since the twentieth century, at 2.3% (as shown in Fig. 3.1).

Even in the case of fluctuating economic growth rate, the rapid development of the tertiary industry has been a major bright spot in the development of China's

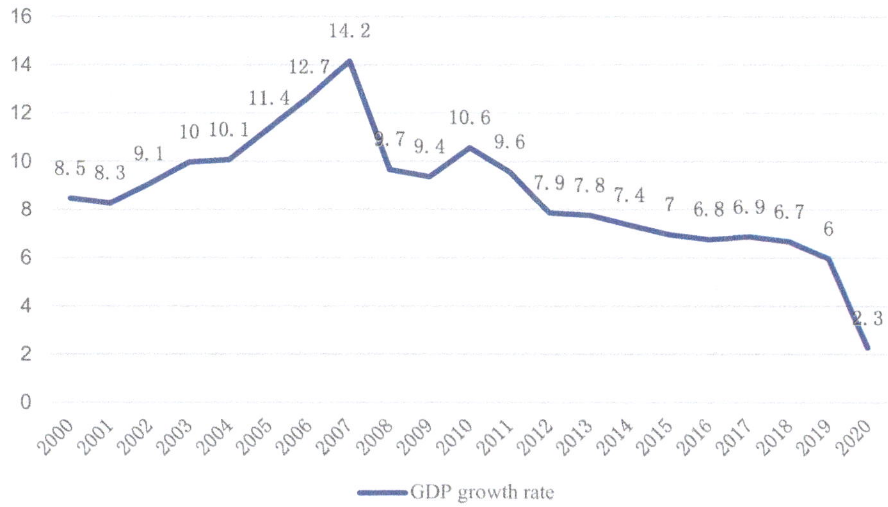

Fig. 3.1 GDP Growth rate from 2000 to 2020. *Source* China Statistical Yearbook from 2000 to 2020

industrial structure since the twentieth century. As shown in Fig. 3.2, the proportion of the tertiary industry in GDP in 2000 was only 39.79%. In the following 20 years, the proportion of the tertiary industry in GDP has been increasing and exceeded the 50% mark in 2015. When the economic growth slowed down in 2014, the proportion of the tertiary industry in GDP was still 48.27%, an increase of 2% points over the previous year. The share of primary industry in the national economy dropped from 14.68% in 2000 to 7.65% in 2020, and that of secondary industry dropped from 45.54% in 2000 to 37.82% in 2020. The growth and development of tertiary industry is the most distinctive feature of the continuous optimization of China's industrial structure in the past more than 30 years of reform and opening up. China has attached great importance to adjusting the proportion of agriculture, light industry and heavy industry, and vigorously promotes the development of tertiary industry. The industrial structure is constantly developing in the direction of optimization and transformation.

3.4.2 *The Structure of Employment*

The structural changes of the three industries are also reflected in the structural changes of the employee. As shown in Fig. 3.3, the proportion of employed workers in the three industries was 50:22.5:27.5 in 2000 and changed into 23.6:28.7:47.7 in 2020. The employee number in the primary industry decreased by 52.8%, the number in the secondary industry increased by 27.6% and the proportion of employed persons in the tertiary industry increased significantly, increasing by 73.5% compared with

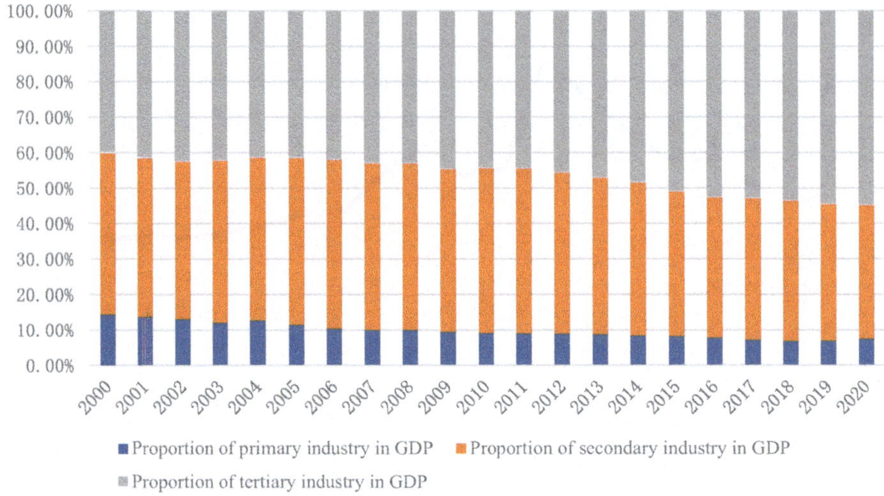

Fig. 3.2 Proportion of GDP of three major industries from 2000 to 2020. *Source* China Statistical Yearbook from 2000 to 2020

Fig. 3.3 Proportion of employee in the three industries from 2000 to 2020. *Source* China Statistical Yearbook from 2000 to 2020

2000. Although the employment structure has changed, it still lags behind the adjustment of industrial structure. The proportion of output value of the three industries in 2020 was 7.65:37.83:54.53. By comparing the proportion of employment in the three industries in 2020, it is not difficult to find that the labor structure deviates from the adjustment of industrial structure.

Low labor productivity in the primary industry led to excessive labor retention. In order to increase labor productivity in primary industries, it is necessary to transfer stranded labor from primary industries, which means a further increase in national output and the corresponding demand for consumption and investment. Recent years, the tertiary industry has become the main sector to absorb employment. Because China's tertiary industry has the characteristics of labor-intensive and low industry access threshold, which can absorb more workers. However, the labor productivity of the tertiary industry is at a low level and the underlying reasons are still the low technical content, the overall quality of employees and the low added value of the industry in China's tertiary industry.

3.4.3 Traditional Industrial Clusters and Emerging Industries

Through decades of large-scale investment and construction, the production capacity and infrastructure service level of basic industries, used to be the "bottleneck" of China's national economy, have been greatly improved. China's energy, transportation, communications and other basic industries have made remarkable achievements. In 2019, China's total energy consumption was 4.88 billion tons of standard coal, an increase of 3.31% over the previous year. In the same period, China's total primary energy production was 3.97 billion tons of standard coal, with an energy self-sufficiency rate of 81.5%, ensuring China's economic security. China's total railway mileage in 2020 reached 146,300 km, an increase of 4.57% from the previous year and 2.13 times that of 2000, ranking first in the world. In 2020, the total length of scheduled civil flights reached 9.43 million km, 6.27 times that of 2000. The cargo throughput of coastal ports realized 8.48 billion tons, 7.55 times that of 2000. The post and telecommunications industry developed rapidly, with the total amount of postal and telecommunications services rising from 23.28 billion yuan in 2000 to 2105.32 billion yuan in 2020, 90.43 times that of 2000. In addition, the Internet penetration rate reached 70.4% in 2020, with 988.99 million people accessing the Internet.

In recent years, China's infrastructure construction has gradually improved and basic industries have developed rapidly. The structure of production and supply is also developing in a people-oriented, energy-saving and low-carbon direction, which can be verified in fixed assets. In 2020, the speed of fixed asset investment in China's heavy and chemical industry slowed down significantly. The fixed asset investment in mining industry decreased by 14.1% compared with the previous year and the fixed asset investment in manufacturing industry decreased by 2.2%. Among them, the fixed asset investment in metal products, machinery and equipment repair industry decreased by 31.3% compared with the previous year.

While the growth of traditional industrial clusters represented by heavy and chemical industry slowing down, the emerging industrial clusters such as high-end manufacturing and modern service industry are developing rapidly. In 2020, high-end manufacturing industries realized a high growth rate of fixed asset investment. Among them, the medical manufacturing industry got a 28.4% increase in fixed asset investment, the computer, communication and other electronic equipment manufacturing industry saw a 12.5% increase. In 2020, the sales revenue of new products in the pharmaceutical manufacturing industry reached 769.81 billion yuan. Electronic and communication equipment manufacturing achieved 4.78 billion yuan, and that of computer and office equipment manufacturing reached 734.37 billion yuan. In general, high-end manufacturing showed a good trend of development in both fixed asset investment and new product sales.

References

Becker, S. Gary., and Kevin M. Murphy. 1992. The Division of Labor, Coordination Costs, and Knowledge. *The Quarterly Journal of Economics* 107 (4): 1137.

Capello, Roberta. 2008. Regional Economics in its 1950s: Recent Theoretical Directions and Future Challenges. *Annals of Regional Science* 42 (4): 747–767.

Dixit, A.K. 1976. *The Theory of Equilibrium Growth*. Oxford University Press.

Fan, Jianyong and Lingyun Shi. 2009. The Externality of Industries, the Competitive Environment of Enterprises, and the Productivity. *Management World* (8): 65–72, 187.

Florida, R. 2002. Bohemia and Economic Geography. *Journal of Economic Geography* 2 (1): 55–71.

Frenken, K., and Ron Boschma. 2007. A Theoretical Framework for Evolutionary Economic Geography: Industrial Dynamics and Urban Growth as a Branching Process. *Journal of Economic Geography* 7 (5): 635–649.

Frenken, K., Frank G. Van Oort, and Thijs Verburg. 2007. Relate Variety, Unrelated Variety and Regional Economic Growth. *Regional Studies* 41 (5): 685–697.

Fu, Linghui. 2010. An Empirical Research on Industry Structure and Economic Growth. *Statistical Research* (8): 79–81.

Gan, Chunhui and Ruogu Zheng. 2009. An Empirical Study on Change of Industrial Structure and Productivity Growth Since the Reform and Opening-up-a Test for the Structure-Bonus Hypotheses from 1978 to 2007 in China. *China Industrial Economics*, (2): 55-65.

Gereffi, Gary. 1989. Rethinking Development Theory: Insights from East Asia and Latin America. *Sociological Forum* 4 (4): 505–533.

Glaeser, Andreas. 2003. Power/Knowledge Failure: Epistemic Practices and Ideologies of the Secret Police in Former East Germany. *Social Analysis* 47 (1): 10–26.

Greunz, Lydia. 2004. Industrial Structure and Innovation-Evidence from European Regions. *Journal of Evolutionary Economics* 14 (5): 563–592.

He, Dexu and Zhanqi Yao. 2008. Effects of China's Industrial Structure Adjustment, Object of Industrial Optimization and Policy Supporting System. *China Industrial Economics* (5): 46–56.

Henderson, Vernon. 1997. Externalities and Industrial Development. *Journal of Urban Economics* 42 (3): 449–449.

Hu, Shuguang, Zhigao Liu and Ying Fan. 2011. Review on the Theory of Industrial Structure Evolution. *Journal of China University of Geosciences (Social Sciences Edition)* (1): 29–34.

Huang, Maoxing, and Junjun Li. 2009. Technology Choice, Upgrade of Industrial Structure and Economic Growth. *Economic Research Journal* 44 (7): 143–151.

Kojima, Kiyoshi. 1986. Japanese-Style Direct Foreign Investment. *Japanese Economy* 14 (3): 52–82.

Krugman, Paul. 1986. *Strategic Trade Policy and the New International Economics*. MIT Press.

Kuznets, Simon. 1967. Quantitative Aspects of the Economic Growth of Nations: X. Level and Structure of Foreign Trade: Long-Term Trends. *Economic Development and Cultural Change* 15: 1.

Kuznets, Simon. 1975. *Seasonal Variations in Industry and Trade*. Arno Press.

Li, Jinyan and Deyong Song. 2008. Specialization, Diversity and Urban Congregation Economy. *Management World* (2): 25–34.

Liu, Wei, and Hui Zhang. 2008. Structural Change and Technical Advance in China's Economic Growth. *Economic Research Journal* 43 (11): 4–15.

Liu, Wei and Shaorong Li. 2002. Industrial Structure and Economic Growth. *China Industrial Economics* (5): 14–21.

Liu, Xiaohua and Tie Lv. 2010. Research on the Characteristics and Policy Orientation of Strategic Emerging Industries. *Scientific Management Research* (9): 20–26.

Liu, Yuanchun. 2003. Economic Institutional Reform or Industrial Structural Promotion—On Core Factor of Chinese Economic Growth and Focus of the Future Reform. *China Industrial Economics* (9): 5–13.

MacKay, D.I. 1969. Industrial Structure and Regional Economic Growth-A Further Comment. *Scottish Journal of Political Economy* 16: 99.

Mao, Fengfu and Jiashun Pan. 2012. Capital Deepening, Industry Structure and Labor Productivity in Chinese Cities. *China Industrial Economics* (10): 32–44.

Nicholas, Kinnie. 1985. Local Managers' Control Over Industrial Relations: Myth and Reality. *Personnel Review* 14 (2): 2–10.

Pasinetti, L. Luigi. 1984–1985. The Difficulty and Yet the Necessity, of Aiming at Full Employment: A Comment Involuntary Unemployment in the Long Run: Pasinetti's Formulation of the Keynesian Argument—A Review Article. *Journal of Post Keynesian Economics* 7 (2): 246–248.

Peneder, Michael. 2010. Technological Regimes and the Variety of Innovation Behavior: Creating Integrated Taxonomies of Firms and Sectors. *Research Policy* 39 (3): 323–334.

Porter, Philip. 2003. Varieties of Capitalism: The Institutional Foundations of Comparative Advantage. *Academy of Management Review* 28: 515–517.

Qiao, Xiaonan, Pengcheng Wang and Jiayuan Wang. 2012. Crossing the "Middle Income Trap": Experience and Countermeasures: A Perspective Based on Marxist Economics. *China Review of Political Economy* (3): 168–184.

Rosenthal, S. Stuart., and William C. Strange. 2001. The Determinants of Agglomeration. *Journal of Urban Economics* 50 (2): 191.

Rostow, W.W. 1960. *The Stages of Economic Growth*. University Press.

Sun, Xiaohua. 2020. Theoretical Reflection and Practical Examination on the "Petty-Clark Theorem"—A Case Study of the Industrial Development and Structural Evolution in India. *Contemporary Economic Research* (3): 47–54.

Timmer, M.P., and A. Szirmai. 2000. Productivity Growth in Asian Manufacturing: The Structural Bonus Hypothesis Examined. *Structural Change and Economic Dynamics* 11 (4): 371–392.

Zhou, Qin and Lihua Wu. 2008. Industrial Structure, Industrial Competitiveness and Regional Employment Differences. *The Journal of World Economy* (1): 78–89.

Chapter 4
Emerging Industries Under Globalization

This chapter introduces the perspective of globalization to analyze the industrialization process of major economies and the development of emerging industries. In the past, globalization was dominated by developed capitalist countries and the United States became the biggest beneficiary of globalization. However, with the progress of technology and the further advance of globalization, new globalization is forming and gradually affecting the global industrial layout. The new round of technological revolution promotes the intellectualization of industrial development and the global labor division and integration. The development of developing countries also changes the attitude of developed countries towards the new round of globalization. As the new round of globalization deepens day by day, the industries of major economies around the world are also transforming from extensive industries to clean and low-carbon industries. Both developed and developing economies have incorporated innovation into their national policy system and governments are playing a leading role in this process.

4.1 Global Changes and New Developments in Globalization

In the long history of human development, globalization is an uninterrupted wave. Globalization has profoundly changed the development trajectory of human society. The great geographical discovery at the end of the fifteenth century connected the world as a whole for the first time, and the opening of new sea routes broke the previous isolation between various regions of the world. Since then, human society has started to embark on the integration process of overall development from scattered growth. If it is regarded as the logical starting point of globalization, the development of globalization has gone through more than 500 years.

Since the beginning of globalization, the leading force has been capitalism. Globalization originated from the great discoveries of geography, namely a series of

Y. Jiang and J. Gu, *Technology and Industrial Transformation of China*,
https://doi.org/10.1007/978-981-19-7458-8_4

important discoveries made by European navigators in the fifteenth to seventeenth centuries about the New World of America and new routes between the east and west. Marx and Engels pointed out in the Communist Manifesto, "The discovery of America, the voyage around Africa, opened up new horizons for the emerging bourgeoisie. Great industry established the world market prepared by the discovery of America. World markets have led to tremendous growth in commerce, navigation and land transportation. This development, in turn, promoted the expansion of industry. The bourgeoisie, opening up the world market, has made production and consumption in all countries cosmopolitan." It is under this circumstance that world trade has achieved tremendous development, and it has promoted the development of social productive forces and economic exchanges. The development of economic exchanges between different regions has pushed human history into the era of globalization. In the development of capitalism, the depth of world history has been enhanced and the process of globalization has been advancing. More and more countries have been involved in globalization with the global expansion of capitalism. It can be said that the rapid development of contemporary globalization is a new stage of "world history" initiated by the expansion of capital and the development of capitalist market.

After World War II, the United States became an important leader and promoter of globalization. However, due to the coexistence of two major world markets, the globalization at that time showed obvious regional characteristics. Many countries began to accept the market economy model as the goal of reform. As a result, globalization spread around the world and the process of globalization was greatly accelerated. During this period, the United States greatly promoted the transnational flow of capital, trade and services, and thus allocated global resources.

The United States has gained great benefits in the process of globalization, which has greatly promoted its economic prosperity and development. The United States has been the biggest beneficiary of globalization. With the continuous development of globalization, the ideology of the United States has been widely spread around the world and its international influence has been greatly enhanced. However, many global problems arose at the same time, such as the escalation of immigration, intensified trade conflicts, rampant transnational crimes and so on. In recent years, rather than sharing in the benefits of globalization, the American middle class has become a victim of it, leading to a surge in populism. As a result, the United States changed its attitude towards globalization and began to adopt anti-globalization policy, from an active promoter of globalization to a promoter of anti-globalization. The US government has taken an anti-globalization stance and refused to provide public goods for global governance, which further divided and reorganized the international system and even endangered the international order based on liberalism and globalism. As the United States is still in a dominant position in the current world system, consolidating and maintaining hegemony is its core national interest. It needs to maintain absolute dominance over countries that can challenge its hegemony or have the potential to challenge its hegemony. But maintaining hegemony and assuming hegemonic obligations is extremely costly. The United States has chosen to reduce its obligations and asked its Allies to bear more of the costs. The most prominent manifestation was the American government's "America First" policy, which has led to a conservative realism in the global economic order. The US's pursuit of maximization of economic interests has led to escalating trade frictions among major economies, posing more challenges to the already troubled international economic coordination mechanism.

The trend of globalization is irreversible. The process of globalization will not slow down or stop for some western countries adopting anti-globalization positions or policies. Anti-globalization does not mean the end of globalization, nor can it prevent the further development of globalization. In the process of the new globalization, the leading power is bound to change, that is, the new globalization will be joined by all countries in the world.

4.2 The Characteristics of the New Globalization

In the process of new globalization, the logic of continuous expansion of globalization driven by global capital flow changed. To a large extent, traditional globalization can be regarded as the gradual development of capitalism on a global scale and capitalism has been the dominant force for quite a long time. However, the rise of new globalization certainly changed the expansion logic that the global capital flow promotes the process of globalization. Globalization led by capitalism, with obvious imprint of capitalism, existed and was accompanied by many major maladies in the process of capital expansion. At the same time, economic globalization has also globalized the negative aspects of the market economy, such as environmental pollution, migration, disease transmission and ecological destruction, which has resulted in confusion and disharmony in the world. At present, the world is undergoing a transformation from the old globalization to the new globalization, where the latter presents a new development trend. The new globalization means more equal participation, more inclusive development and more shared benefits.

4.2.1 New Industrial Revolution and Integration

Technological progress is one of the key driving forces of globalization. The promotion of technological innovation has two aspects. First, new inventions improve the speed of transportation and communication and reduce the trade cost. Second, technological progress improves the production mode, increased tradable goods, improved production efficiency and changes the direction of trade. The progress of information technology and the improvement of global informatization have greatly reduced the cost of information transmission and formed a new paradigm of globalization based on task division and trade. At the bottom line, it is about the impact of information technology on reducing the trade cost. At present, the information technology revolution is still upgrading, and its influence on globalization is also continuing and deepening. As mentioned above, the industrial revolution based on the new generation of information technology is emerging in various countries led by industrial Internet, Industry 4.0 and Made in China 2025. New manufacturing industries such as 3D printing continue to merge, and production is adjusted from the 3.0 mode of automation to the 4.0 mode of intelligence. New consumption modes such as platform consumption and sharing economy emerge endlessly, and product design, production, exchange and consumption are increasingly integrated. Innovation and convergence not only improve manufacturing efficiency, but also change the mode of production and business competition.

First, the technological revolution promotes intelligent production. Based on the Internet and big data, intelligent production has improved the feasibility of mass customization to meet the personalized needs of the market. The connection between supply and demand is more rapid and direct, and the manufacturing efficiency is greatly improved. As a result, it may change the production processes, reduce workflows, and even affect enterprise location and industrial agglomeration. The emergence of new production methods such as 3D printing shortens the production chain. These changes will affect the pattern of international production and labor division.

Second, the production and business model changed, and the labor division and trade based on enterprise functional modules may become a new paradigm of globalization. In the wave of new technological and industrial revolution, the development and production of new products and services, such as general Electric's cloud service platform, Tesla and Uber, which represent the deep integration of production and manufacturing with IT technology and Internet, adopt modular production organization. The modular organization has its roots in the Silicon Valley's computer industry. In the modular way, the core enterprise no longer complete product design, production and sales in one package, but constantly innovate the standardized interface of system and product design by a large number of module suppliers with advantages in the segmented field independently design and produce shared professional modules, and then the core enterprise completes product integration. In the era of new paradigm of globalization, multinational manufacturing companies are still vertically integrated production organizations, and organize production and sales globally based on their monopolistic advantages in product technology and production. At present, the modularization of the production organization mode of machinery, automobile and other manufacturing industries and service industries means that the production and business models of these industries will change in the future. Enterprises compete in the standardization of interface, the integration of modules, the functionalization and specialization of modules through standard sharing and horizontal cooperation. Eventually, an industrial pattern composed of a small number of system designers or integrators and a large group of modules may be formed. In this pattern, R&D and design, internal functions of enterprises, services, products and components production are likely to become independent modular tasks. Therefore, the functions and production activities of enterprises are subdivided into specific tradable tasks, which will be divided into division, exchange and integration on a global scale. The paradigm of globalization may change into a new paradigm of module division of labor and trade, which is different from the new paradigm based on task division and trade, and become a new paradigm of globalization. Under the new paradigm of globalization, the degree of system or platform computing and networking, intelligent machines and function division level become the key industry competitiveness. Innovation environment, manufacturing process, giant market and resource integration platform may become the absolute or comparative advantage of a country, an enterprise and even an individual. Therefore, the global labor division pattern is constantly being changed and adjusted.

4.2.2 The Attitude of Developed Countries

The developed countries led by the United States changed their positive attitude towards globalization to being cautious and even resistant, which is based on two aspects of reflection.

One is for these countries to reflect on their own development model. Faced with sluggish economic recovery after several expansionary policies, developed countries began to rethink their past development models and realize the significance of manufacturing for employment and economic expansion. The United States, Germany and other developed countries have formulated re-industrialization strategies. The United States issued the Framework for Revitalizing American Manufacturing at the end of 2009, launched the Advanced Manufacturing Partnership program in 2011, proposed the construction of a National Manufacturing Innovation Network in 2012, and released the Preliminary Design of the National Manufacturing Innovation Network in early 2013, gradually implementing the strategy of "re-industrialization" and "Revitalizing American manufacturing". Germany released high-tech Strategy 2020 in 2010 and put forward the "Industry 4.0" strategy. In 2014, the Digital Agenda (2014–2017) was issued to promote digital innovation and the digital society construction. These re-industrialization strategies aimed to revitalize manufacturing through a new industrial revolution. In recent years, General Electric Company (GE), IBM, Google, Siemens and other industry giants have begun to move to actively promote the integration of industry with big data and the Internet, and promote the informatization and intelligence of industrial production. The process of new industrial revolution may drive industrial expansion in developed countries. After domestic production replacing some offshore outsourcing to a certain extent, the global value chain dominated by developed countries shrank to some extent. In this case, the impact of the contraction of GVCS would be long-term.

Another is a reflection on the benefits of globalization. In view of the domestic dissatisfaction with globalization, developed countries began to rethink the benefits of globalization, and there was an anti-globalization trend. The US government put forward the idea of strengthening the national first doctrine and re-examined the agreements signed or negotiated before that had been aimed to promote global trade or regional trade, which leads to withdrawing from TPP (Trans-Pacific Partnership) negotiations, renegotiating NAFTA (North American Free Trade Agreement) and proposing to revise the US-South Korea Free Trade Agreement. In August 2017, US

Trade Representative Robert Lighthizer announced the launch of section 301 investigation against China, indicating that the US has begun to implement trade protection measures. The United States has also taken measures to limit immigration or immigration of certain ethnic or religions groups. Similar anti-globalization actions or measures have emerged in Europe. For example, the UK voted to leave the EU in the referendum in 2016, the rise of far-right parties in Europe, and Germany approved new regulations to strengthen the review of acquisitions of strategic enterprises by non-EU enterprises. These anti-globalization measures restricted to varying degrees the free movement of goods, investment and people, and hindered the expansion of globalization for some time. But the effect will not last. International economic theory suggests that domestic measures such as aid for trade, rather than restrictions on free trade, are usually and effectively used to counter the negative effects of free trade and globalization on income distribution. New globalization has certainly played a role in the widening income gap within the United States, but technological progress, not international trade, has been the main factor. Anti-globalization measures such as trade protection cannot solve the fundamental problem. In the future, the US will return to its usual practices such as trade adjustment assistance measures. On this basis, it also needs to adjust its competition strategy, education strategy, welfare policy and industrial policy.

4.2.3 China as a Major Promoter of Globalization

Attention to the adjustment of the global governance system is now focused on China. It is not only because of the remarkable achievements China has made in recent years, but also because of the collective rise of emerging economies and developing countries represented by China's development and growth, which marks a historic transition in global governance from "western governance" to "East-West co-governance". As the world is changing, the global governance system needs to be reformed and changed accordingly.

Globalization needs China and has long been inseparable from China. The size, weight and energy of the Chinese economy in the world economic map are undoubtedly an important support and guarantee for China to promote a new type of economic globalization. Since the founding of the People's Republic of China, especially over the past four decades of reform and opening-up, China has made remarkable economic achievements. It has become the world's second largest economy, the largest industrial country, the largest trader and the largest holder of foreign exchange reserves. After the outbreak of the international financial crisis, the world economy has been in a downturn. The rapid development of Chinese economy has become an important engine of global economic growth, providing a strong momentum and impetus for the steady growth and sustained development of the global economy.

Economic globalization has suffered setbacks. The globalization led by western countries has encountered difficulties, mainly because the existing economic growth model, governance model and development model have exacerbated the imbalance of global development to a certain extent, widening the gap between regions and countries instead of solving the problem. At the same time, as leaders and beneficiaries of globalization, western countries also face serious development-space imbalance and unequal income distribution, leading to severe social polarization and unequal sharing of the benefits of globalization among low and middle-income people. Different from the economic globalization led by the developed capitalist countries in the west, the new type of globalization promoted by China adheres to the principle of seeking success by pursuing shared benefits through consultation and collaboration, and targeting a community with a shared future for mankind. Standing for the high ground of human values and morality, the idea of "a community with a shared future for mankind" transcends the "western version" of globalization, abandons the Western-centric theory, rejects the hegemonic logic and bullying, and rejects the old thinking of the "zero-sum game" and "law of the jungle".

If the "community with a shared future for mankind" is a wonderful blueprint for the new type of economic globalization presented by China from the perspective of values and ideals, the "Belt and Road" initiative reflects China's promotion and implementation of the new type of economic globalization from the perspective of policies and practices. As an important milestone in China's opening-up policy and the implementation of the significant move of all-round opening to the outside world, "One Belt and One Road", from a concept to the blueprint, a plan to practice, the initiative to action, is China's push to build the mutually beneficial and win-win benefit community. As a responsible major country, a socialist country and a developing country, China actively explores and innovates in promoting a new type of globalization. In the spirit of multilateralism, China is promoting an open world economic system while adhering to the principle of extensive consultation, joint contribution and shared benefits. China sticks to its efforts in building a global industrial, value and supply chain that benefits all, overcoming the disadvantages of capital-driven globalization, focusing on the development of the real economy, and making the global economy more inclusive and balanced. Positive progress has been made in the "Belt and Road" initiative, which connects dots into lines and expands from lines to planes. "Belt and Road" initiative has opened up high horizons for China's opening-up and provided new opportunities for the development of other countries. Cooperation in fighting the epidemic has brought new meaning and expectations to the "Belt and Road" Initiative. The "Belt and Road" initiative should be built into a road of cooperation to jointly meet challenges, a road of health to safeguard people's health and safety, a road of recovery to promote economic and social recovery, and a road of growth to unleash development potential.

4.3 Emerging Industries in Developed Economies

The global financial crisis has driven the adjustment and change of the international industrial structure. Major economies in the world are taking the competition for the commanding heights of science and technology as a national strategic priority and vigorously developing emerging industries. In the process of deepening globalization, developed economies, according to the law of the development of emerging industries, reasonably choose the key links and fields of industrial development, carry out corresponding policy support, and keep reform and innovation to maintain their advantages in the globalization.

4.3.1 America and Technological Edge

4.3.1.1 Making Technological Innovation and Emerging Industries Development as National Strategies

The US government has always attached great importance to technological innovation. Since the Obama administration, innovation has been regarded as the key to readjusting the US economic structure and restoring the vitality of the US economy. In September 2009, the United States launched the National Innovation Strategy "Advancing Sustainable Growth and Quality Jobs". In February 2011, a new version of the National Innovation Strategy "Ensuring Economic Prosperity and Growth", argues that America's future economic growth and international competitiveness depend on its ability to innovate and only by vigorously developing new technologies can the US economy achieve rapid and sustainable growth and provide more high-paying jobs. To achieve this goal, the U.S. public and private sectors must work together to develop education, basic research, information technology, infrastructure, and create a favorable environment for technological development and industrial innovation. The new version of the National Innovation Strategy identifies clean energy, biotechnology, nanotechnology, advanced manufacturing, space technology, health care technology and education technology as priority direction to promote technological innovation and the development of emerging industries.

4.3.1.2 Investment in R&D Is Increasing

While the 2008 financial crisis hit the U.S. economy hard, it didn't have much impact on the overall level of R&D cost in the United States. Research and development funding in the United States comes mainly from corporations, the federal government, universities and other non-profit organizations. Enterprises are the main providers of R&D funds and also the main undertaker of R&D activities. The R&D input provided by the federal government is mainly used for basic research, generic technology research and R&D equipment. During his inauguration, the Obama administration proposed to increase R&D investment to 3% of GDP in order to further promote technological innovation and the development of emerging industries in the United States. From 2009 to 2012, US R&D spending as a share of GDP has remained stable at around 2.8%, and its share of global R&D spending has remained at around 32%. In 2017, the US still spent 2.79% of its GDP on R&D. For its wonderful research and development environment and abundant human resources, the US has also attracted a large number of foreign companies to set up research and development institutions in the US, which has also played a positive role in the development of new technologies in the US.

4.3.1.3 Clean Energy as the Development Direction of Emerging Industries

The United States has put the technological breakthrough and industrial development of the new energy industry in a very prominent position, expecting to lead the new clean energy industry revolution by vigorously developing top clean energy technologies. This development direction can not only maintain the leading position of the United States in the forefront of technology and achieve industrial development and employment increasing, but also reduce the dependence on overseas energy, so as to ensure the energy security of the United States and realize the goal of energy conservation and emission reduction of greenhouse gas emissions. According to the Statistical Yearbook of World Energy, the US emitted 445.72 million tons of carbon dioxide in 2020, down 11.6% from the previous year, showing initial results in energy conservation and emission reduction.

In the American Recovery and Reinvestment Act of 2009 (ARRA), the United States has invested US $90 billion in the development and utilization of clean and efficient energy, such as promoting renewable energy, developing and building transportation and high-speed railway, smart grid technology, tax deduction for the production of clean energy equipment. The US government and Congress were engaged in a continuous game on the government debt issue. Even in the case of financial difficulties, the U.S. government has been providing financial support for new energy within its capacity. In the US 2016 federal budget, $6.7 billion was allocated for research, development, demonstration and deployment of new energy sources. The 2017 budget increased by about 30% over the previous year to continue supporting the development of new energy technologies, including advanced bio-energy and emerging nuclear energy technologies. The United States led the world in nuclear energy consumption, reaching 7.39 EJ in 2020, accounting for 30.8% of the global total. Renewable energy consumption reached 6.15 EJ, accounting for 19.4% of the world's total consumption. It can be seen that the United States of America's massive investment in the development and commercial application of new energy technology has achieved obvious results.

4.3.2 European Countries and Industrial Development

4.3.2.1 "Innovation" and "Green" as the Focus of Development Strategy

In order to revive the European economy from the financial crisis and continue to be well positioned in the global competition in the future, the European Commission issued the Europe 2020 Strategy in 2010, which set out the development goals and strategic priorities of the European economy in the next 10 years. The EU put "innovation" and "green" in a prominent position, hoping to establish the overall competitiveness based on green and low-carbon economy through technological and industrial innovation so as to reshape the EU's economic vitality and stimulate employment growth. Therefore, EU countries have promoted innovation and regarded the development of emerging industries as the main way to reshape their competitive advantage. In the report "Building Britain's Future: New Industries and New Jobs", the British government proposed that Britain should focus on developing emerging industries such as low-carbon industry, life science and pharmaceutical industry, digital industry and advanced manufacturing industry in the future. In order to implement this strategy, the UK implemented the "Technology and Innovation Center Plan" in 2011. Up to now, five innovation centers have been established, including manufacturing technology, marine renewable energy technology, renewable medicine technology, satellite application technology and digital economy technology, providing important support for the commercial development of products and services in relevant fields in the UK. The German Government has adopted the German High Technology Strategy 2020, which identified five key areas for development including climate and energy, health and nutrition, transportation, safety and communications, and proposed a series of policy measures to promote high-tech development.

4.3.2.2 Developing Low-Carbon Economy as the Core of Scientific and Technological Research

The term "low carbon economy" first appeared in government documents in UK's 2003 energy white paper "Our Energy Future: Building a Low carbon Economy". It emphasized scientific innovation priority to develop renewable energy, including wind, water (wave and tidal power), biomass energy, solar energy, energy crops and established several technical fields most likely to produce substantive breakthroughs such as carbon dioxide absorption, energy efficiency, hydrogen production and storage, nuclear energy (especially waste treatment) and tidal energy. The UK also issued the climate change strategic framework, which put forward the vision of a global low-carbon economy and pointed out that the impact of the low-carbon revolution could be comparable to that of the first industrial revolution. In October 2004, the European Commission announced that it had approved emissions plans for eight EU member states, limiting the allocation of their carbon dioxide emissions. In 2006, the European Commission's Action Plan - Realizing the Potential for Energy Efficiency was launched. This Plan introduced more than 70 actions across the EU over a six-year period, resulting in significant improvements in energy efficiency, which the Commission estimates would reduce energy consumption by 20% by 2020. In addition, the European Commission has published a Roadmap for the Development and Investment of low-carbon Technologies, with a total investment of around 53 billion euros between 2010 and 2020. According to the World Energy Statistical Yearbook, Europe's CO_2 emissions totaled 3596.8 million tons in 2020, down 12.3% from 2019 and 23.1% from 2010.

With the adjustment of the energy structure of European countries, the research and application of new energy in Europe have witnessed great development. In 2020, the total consumption of renewable energy in Europe was 8.94 EJ, an increase of 8.2% over the previous year, among which Germany's consumption of renewable energy was 2.21 EJ and Britain's was 1.2 EJ, an increase of 9.9% over the previous year, ranking among the top European countries. In terms of renewable capacity, the UK generated 127.8 TWH of electricity in 2020, up 11.2% from the previous year. Germany's renewable energy capacity is the highest in Europe, reaching 232.4 TWH in 2020, up 4.1% from the previous year. According to the data provided by the Statistical Yearbook of Energy, the total installed capacity of solar energy in Europe in 2020 reached 167.8 GW, accounting for 23.7% of the global total. Among them, the installed capacity of solar energy in Germany ranks the first in Europe with 53.8 GW, and Italy ranks the second with 21.6 GW. According to the European Wind Energy Association, Europe had 216.6 gigawatts of installed wind capacity in 2020, accounting for 29.5% of the global total. Germany had the highest installed wind capacity in Europe in 2020 with 62.2 GW, while Spain was second with 27.1 GW. In terms of biofuel production and consumption, Europe's share of the global market has also exceeded 20%, realizing the research and development and application of new energy technologies.

4.3.3 Japan and New Growth Engine

In the economic development history of all countries in the world, Japan is a typical late coming one, that is, a typical late developed industrial country. Although Japan's surpass is the result of many factors, there is a close internal relationship between the transformation of its identity and the development of manufacturing industry. As a country lacking of energy resources, especially energy, the important reasons why Japan could successfully overcome the environmental costs of the serious oil crisis and achieve rapid growth in the 1970s were the transformation of its manufacturing industry in this period by replacing resources with technology and the transformation from resource-consuming heavy chemical industry to knowledge-intensive industry.

It is precisely because of the importance of manufacturing industry to Japan's economic development that the Japanese Industrial Revitalization Plan, one of the important action frameworks of the "economic growth strategy" of Abe-economics' third arrow, was adopted by the Japanese cabinet meeting on June 14, 2013. It involved the urgent structural reform centered on promoting private investment, the reform of employment system and the cultivation of human capital, the promotion of technological innovation aimed at competing for the commanding heights of science and technology, and the construction of the world's highest level information society. In recent years, Japan's reform of manufacturing industry mainly consisted of the following directions.

4.3.3.1 Developing the Robot Industry

Japan stands in a leading position of robot industry whether in in terms of technology, production and supply of industrial robots. So far, Japan has maintained the world's largest number of robot shipments and operating units. It accounts for more than 90% of the world's production and supply of precision retarders, servomotors, sensors and main components of robots. At the same time, Japan also has a world-class level in the research and development of humanoid and service robots. In January 2015, Japan formulated the new robot strategy to revitalize the Japanese economy. This strategy brought Japan back to the forefront of the world economy. Japan has realized that today's robot technology and industrial development should adapt to the direction of world development in the future. In order to comply with the trend of development, its original technology and system must be transformed. Japan has proposed the task of robot revolution, which was easy to use and suitable for the needs of diverse and multiple fields. The integration of robot and information network technology has become a new source of added value creation in the field of manufacturing and service. It has realized the "robot barrier-free society" in which humans and robots coexist. At the same time, Japan conducted field research on the technological innovation activities of Japanese robot manufacturing enterprises such as Yaskawa motor.

4.3.3.2 Reviving Small and Medium-Sized Enterprises as an Impetus to Development

The 3.85 million small and medium-sized enterprises account for 99.7% of the total number of Japanese enterprises, which have always been the foundation of Japan's economic development. In particular, small and medium-sized manufacturing enterprises created more than half of the added value of the manufacturing industry, which supported the development of the Japanese economy. In recent years, the Japanese government has constantly formulated various laws and policy measures to support and foster small and medium-sized enterprises. In order to cope with the changes of economic situation and domestic and international economic environment, solve the problem of weakening overseas production, and make research-oriented enterprises better adapt to the needs of market and related enterprises, the Japanese government formulated relevant laws based on the advanced manufacturing technology of small and medium-sized enterprises. The relevant institutions of the economy ministry have systematically sorted out the research directions of various basic manufacturing technologies that can strengthen competitiveness and help create new products, formulated the guidelines on the upgrading of specific manufacturing basic technologies for small and medium-sized enterprises, and continuously adjusted and revised them. In addition to financial and fiscal support measures in accordance with the law, the government has also adopted various forms to encourage and promote innovation, such as selecting local promising small and medium-sized enterprises as the benchmark.

4.3.3.3 Reallocating the Roles of Manufacturing Enterprises at Home and Abroad

Japan's resources and market conditions in terms of economic development made it unable to change its extroverted characteristics in economic development. In the process of deepening economic globalization, Japan still takes the world market as its main stage of economic activities. Nikkei reported in 2017 that some Japanese companies would move back to Japan from overseas markets, mainly in the home appliances, precision machinery and automobile industries. Panasonic, for example, moved microwave ovens, air conditioners and washing machines made in China back to Japan. SHARP moved its production of air purifiers, among other things, back to Japan from its Shanghai base. At the same time, Japanese companies were accelerating domestic investment. Honda has built a world class energy efficient plant in Omari, Saitama prefecture, to produce environmentally friendly products. Olympus decided to invest 19.7 billion yen to expand its endoscope production capacity by 30% by 2017. Japanese monitors are invested in Shiguyuan factory in Chiba County, in which produced high-performance and high-quality smart phones and tablet computer monitors with about 200 billion yen of equipment investment.

4.4 Industrialization of Developing Countries

With the deepening of globalization, emerging economies represented by BRICS countries are in a new round of industrialization. Although different countries are at different stages of economic development, they will face different factor endowments, resource constraints and market environment, thus choosing different economic development models. Brazil's industrialization process is of typical significance. This paper analyzes the road of Brazil's industrialization, so that we can have a deeper understanding of the important impact of industrialization development on the rise of emerging powers. In addition, Russia, South Africa, India and China, as

BRICS countries, have also made significant progress in industrial development. Due to the difference in factor endowment among different countries, they have different paths to develop leading industries.

4.4.1 The Industrialization Process and Industrial Transition in Brazil

Brazil's industrialization process began in the 1930s, with import substitution-oriented industrialization at the beginning. As the largest country in Latin America, Brazil is rich in mineral resources. For a long time, Brazil exported a lot of primary products to western countries in exchange for industrial goods. The export of primary products did not promote the industrial sector, but it marked the industrialization process of Brazil with import substitution. During Kubitschek's administration, Brazil's economy has grown rapidly, including infrastructure improvements with more than 11,000 km of new roads and electricity generation rising from 3 to 5 GW. The heavy and chemical industries have also developed significantly, with Brazil's steel production reaching 2 million tons a year. The rapid development of agricultural sector has also brought financial support and market support for industrialization. Fertilizer production far exceeded the projected with 33%, and domestic agricultural tractors began to replace imported products on the market.

Although the import substitution strategy has brought the prosperity of Brazil's economy, there were also insurmountable disadvantages. As commodity prices continued to decline, Brazil's terms of trade were deteriorating. Domestic savings were insufficient to support the massive investment in industrialization, which had to be offset by external borrowing, resulting in a threefold increase in Brazil's external debt. In the end, the military government suppressed Brazil's inflation by means of coercion, including import substitution and special export substitution, and positive attitude towards foreign loans, which also laid hidden dangers for the subsequent debt crisis. In the 1970s, the financing gap of Brazil's industrialization widened and official loans were unable to meet domestic demand, so the Brazilian government turned to commercial loans in the international financial market. With the intensification of the spillover effect of the monetary policy of the United States and the reform of the international financial market, Brazil could not bear the huge debt, eventually debt crisis occurred.

Along with the deepening of the globalization, the world economy had experienced a boom period. The prosperity of the world economy inevitably promoted the expansion of global trade. The rising demand of the international market has greatly improved the Brazil's terms of trade. A large number of export primary products promoted the sustained growth of Brazil's economy. Brazil's economy has grown at an average annual rate of 3.75% between 2000 and 2008, reversing the slump of the 1980s. The export boom in the early twenty-first century provided a rare opportunity for Brazil to revitalize its manufacturing industry. During this period, Brazil's industrial structure was also changing. In 1949, when Brazil started industrialization, agriculture accounted for 25.97% of GDP, while industry accounted for 26.2%. With the acceleration of industrialization, the proportion of agriculture in GDP declined rapidly. In 1989, the proportion of agriculture in GDP dropped to less than 10%, and in 1995, it dropped to about 5%. Manufacturing is the main part of the industrial sector. From the 1960s to the mid-1980s, the proportion of manufacturing in the industrial sector in Brazil has remained at about 75%.

At the beginning of the twentieth century, in order to reduce dependence on overseas energy, Brazil took advantage of its own advantages and vigorously developed the new energy industry focusing on biomass energy. Since 2010, Brazil's renewable energy generation has increased year by year, from 34.1 TWH in 2010 to 117.6 TWH in 2019 and by 2.1% from 2019 to 120.3 TWH in 2020. Solar installed capacity has been increasing year by year since 2016, reaching 7.9 GWH by 2020. Brazil's installed wind capacity rose from 0.9 GWH in 2010 to 14.4% in 2020, accounting for 65.2% of South America's total installed wind capacity. Biofuels are growing fastest among renewables, with Brazil's biofuel production increasing from 292 MBOE in 2010 to 411 MBOE in 2019, accounting for 23.5% of the global total.

4.4.2 The Innovation and Key Technologies Development in Russia

Russia is rich in natural resources and strong in scientific research, but its economic growth is overly dependent on the energy industry. The collapse of oil prices caused by the global financial crisis has made a serious impact on Russia's economy. Russia realized that it should transform from an energy-dependent economy to an innovation-driven one as soon as possible. Russia also accelerated the pace of developing new industries. In December 2011, Russia officially approved the 2020 Innovation Development Strategy, which clearly outlined the goals, priorities and national policies for innovation in Russia until 2020. The goal of the innovative development strategy is to transform Russian economy into an innovative one. In July 2011, Medvedev identified 8 areas and 27 key technologies for the priority development of Russian science and technology, mainly including security and counter-terrorism, information and communication, life science, future high-tech weapons, energy conservation and nuclear technology and other fields.

Russia has made specific plans for the development strategies in these key areas and attached great importance to nano-technology. The national special plan for nano industrial infrastructure of the Russian Federation was established in 2007. The Russian government invested 5 billion rubles (about 170 million dollars) to provide special fund support for equipment procurement and initially established a National Nanotechnology network system. In the life sciences, Russia has increased government support for the pharmaceutical industry. The initial financial investment for 2011–2015 amounted to 6.15 billion rubles (about 200 million dollars). Before

2015, Russia focused on the establishment of pharmaceutical product innovation centers and talent training. From 2015 to 2020, it paid attention to the research and development of new drugs and medical products, and finally won a place in the international pharmaceutical market.

In order to accelerate the transformation of technological achievements from laboratories to products and combination of industries, universities and research institutes, Russia has also set up a number of high-tech parks. What's more, Russia has constructed an innovation research center in Skolkovo, supporting enterprises in five fields, communications technology, biopharmaceutical, space technology, nuclear energy and energy conservation. The Skolkovo Innovation Park has operated well and was known as the "Silicon Valley" of Russia. Russia continued to expand investment in high-tech parks and has built 24 special economic zones and 12 high-tech parks, 670 enterprises engaged in biomedical and nanotechnology have settled in the parks. In the future, Russia will expand the proportion of innovative products in total social output.

4.4.3 Innovation-Driven Industrial Growth in India

Since the 1990s, India's information, pharmaceutical and other emerging industries have sprung up, which has become a landmark event in the process of technological catch-up of emerging economies. India made full use of its unique labor resource endowment, seized the favorable opportunity of economic globalization and transformed its abundant scientific and technological human resource advantage into the competitive advantage of high-tech industry, enabling its emerging industries to achieve rapid development. India regards innovation as the engine of economic prosperity and the source of national competitiveness in the twenty-first century. In order to better promote technological innovation and the development of new industries, the President of India has identified the period from 2010 to 2020 as the Decade of Innovation in India. India has established a National Innovation Council to promote commercialization and large-scale production of innovation results, thus establishing an innovation institutional framework for inclusive development with Indian characteristics.

To better promote the development of the new energy industry, India has reorganized the government agency and renamed the Ministry of New and Renewable Energy, which was responsible for strategic planning, policy formulation and implementation of new energy development in India. Since 2010, India's renewable energy generation has increased year by year after it set clear targets for new energy development. According to the International Statistical Yearbook of Energy, India's renewable energy capacity rose from 33.9 TWH in 2010 to 151.2 TWH in 2020, accounting for 4.8% of the global total. In addition, India provided financial support for various renewable energy projects through the National Clean Energy Fund and the Green Bank. These fiscal and financial policies have attracted a large amount of social funds to invest in the new energy industry. India has also made a huge leap forward in solar capacity, which achieved 0.1 GW in 2010 to 39 GW in 2020, an increase of 11.5% from 2019, accounting for 5.5% of the global total. In terms of installed wind capacity, India increased from 13.2 GW in 2010 to 38.6 GW in 2020, accounting for 5.3% of the global total. India has also become a new energy power in the Asia-Pacific region, second only to China.

With the rapid development of Indian economy in recent years, many multinational companies have realized the great potential of Indian scientific human resources in the advanced technology research and development. The Indian government was also open to multinational companies joining the R&D process. At present, India has more than 200 multinational companies including Intel, Microsoft, Google, Pfizer and other world-famous companies set up R&D centers in India. Thousands of local technicians supported by multinational companies have carried out research and development in the forefront of emerging technologies and achieved a series of innovative results. AMD's Indian R&D center undertakes nearly half of the R&D work of the company's chips. Welch center in Bangalore has become the largest design and development organization of General Electric (GE) outside the United States. India has also made remarkable achievements in the IT technology research and development. With more and more multinational pharmaceutical giants joining in one after another, India has gradually become a global center for clinical trials of new drugs. It is India adheres to innovation can it achieve unprecedented breakthroughs in emerging industries.

4.4.4 Industrial Development in South Africa

South Africa has developed rapidly in science and technology in recent years and ranks among the world's top in some industries. However, South Africa's science and technology foundation is still very weak generally. To promote the development of new technologies and emerging industries in South Africa and lay a solid scientific and technological foundation for sustainable economic development, the Ministry of Science and Technology of South Africa has issued a series of policies to systematically plan the national innovation system and priority R&D areas. In 2002, the South African Ministry of science and Technology issued the South African National Research and development strategy. It put forward three strategic priorities of promoting innovation, promoting the construction of science and technology talent resources development and promoting effective government science and technology management system, which outlined the overall framework for South Africa's scientific research and defined the key research and development fields. On this basis, South Africa constantly developed new strategic plans to further identify key technology areas requiring focused support. South Africa has successively issued the South African nano-technology strategy, biotechnology strategy and space strategy, and implemented a series of supporting policies.

South Africa has the world's fifth-largest coal reserves, but it lacks oil resources. Therefore, South Africa's power structure is dominated by thermal power. With the increasing economic growth, the contradiction between power supply and demand has become more prominent. The extensive use of coal-fired power generation led to environmental pollution and carbon emissions. For encouraging the development of clean energy, South Africa has introduced a series of policy measures such as "Renewable Energy Protection Price", "Renewable Energy Financial Subsidy Scheme" and "South Africa Wind Energy Project". In recent years, South Africa's renewable energy capacity has jumped from 0.5 TWH in 2010 to 12.6 TWH in 2020, accounting for 29.8% of Africa's total. Solar installed capacity has also reached zero, rising from 0.3 GW in 2013 to 5.5 GW in 2020. South Africa's coal generation has also declined, from 218.2 TWH in 2019 to 202 TWH in 2020, as new energy sources continue growing.

4.4.5 Development of Strategic Emerging Industries in China

As the largest developing country in the world, China has made remarkable achievements in its economic construction over the past 40 years of reform and opening up. Different industrial sectors have made remarkable progress in terms of scale, quantity and technological level. At the same time, China's industrial development was characterized by extensive, polluting and low added value. While entering the stage of normal economic development, the question of how to get rid of the problems existing in the previous industrial development and effectively improve the international competitiveness of the industry has become urgent tasks in the process of China's industrial transformation.

Generally, China's traditional industries attached great importance to the expansion of quantity and scale while ignoring the upgrading of quality and technology, which resulted in a series of problems such as overcapacity, resource shortage, low quality and technology level, low energy utilization rate and environmental pollution. In 2006, for example, China accounted for 19% of global alumina consumption, 30% of global energy consumption, 15% of global energy consumption, and only 5.5% of global GDP. Since the middle and late 1990s, China has entered into the middle stage of industrialization and gradually to the accelerated stage of heavy chemical industry. For a long time, China's energy utilization has been dominated by coal. Excessive dependence on fossil energy for economic development inevitably affected energy security. Accelerating consumption of conventional fossil energy reserves also brought serious damage to the ecological environment.

For high energy consumption, China has issued a series of policies and measures to support the development of new energy. In March 2011, the Notice on Further Promoting the Application of Renewable Energy in buildings clearly pointed out that the proportion of renewable energy such as solar energy, shallow geothermal energy and biomass energy in construction energy should be effectively increased. In 2012, China issued the "Twelfth Five-Year" Plan for solar photovoltaic Industry, which further clarified the construction focus and output of photovoltaic. In terms of the specific development content, China's new energy industry has been improved.

At present, technologies such as high-efficiency concentrating solar cells, large-scale grid connected wind power generation and fourth generation nuclear power are making rapid breakthroughs, triggering explosive growth in the new energy industry. China's renewable energy generation reached 863.1 TWH in 2020, up 16% from 2019, and the annual average growth rate from 2009 to 2019 was 31.3%, accounting for 27.4% of global renewable energy generation. China's installed solar capacity has increased year by year since 2010, reaching 253.8 GW in 2020, up 23.7% from 2019 and accounting for 35.9% of the global total. China's installed wind power capacity has also grown rapidly, from 29.6 GW in 2010 to 282 GW in 2020, accounting for 38.5% of the global total. In the development of nuclear energy, China's nuclear safety regulatory authorities constantly revised nuclear safety laws, standards and regulations in an effort to improve nuclear power. China's nuclear energy consumption also increased from 0.7 EJ in 2010 to 3.25 EJ in 2020, accounting for 13.6% of the global total.

China has also accelerated the construction of high-end device manufacturing. The cultivation of high-end manufacturing industry is not only an urgent need to build new international competitive advantages and seize the initiative in development, but also an inherent requirement to transform the mode of economic development and promote the upgrading of industrial structure. From an international perspective, the financial crisis has made industrialized countries reattach importance to the development of real economy and put forward "re-industrialization", aiming at high-end manufacturing fields and emerging industries, seeking to shape new competitive advantages. It not only constituted fierce competition for the future development of high-end equipment in China, but also left market space squeezed for products that have formed advantages. Domestically, the transformation and upgrading of key industries, the cultivation and development of strategic emerging industries and the construction of major national projects have put forward higher requirements for the green, intelligent and service-oriented device manufacturing industry, and provided a huge market demand space. China has achieved preliminary results in various fields of high-end manufacturing industry. Significant progress was made in research, development and production of aviation equipment manufacturing industry. The development of C919 and ARJ has been promoted, while subcontracting production of international parts in China continued to increase and Boeing's domestic procurement of parts has increased by 20% annually. In addition, China's satellite application industry changed from scientific research to application and service development mode. China already has the conditions for the independent construction of space infrastructure. The BDS navigation System has been put into trial operation, and the industrialization of the BDS has entered into the stage of substantial promotion. China's rail transport system has developed into an independent development capability with international influence, represented by "high-speed rail" which has 820 high-speed trains and has reached 9,356 km operation length, ranking first in the world.

During the period of "11th five-year plan", the approval of the State Council issued "a number of policies to promote the development of biological industry to speed up" and "biological industry development of the eleventh five-year plan", vigorously stimulated the industrialization of biotechnology research and development. A large number of innovative enterprises have emerged, and technological innovation capacity has been greatly enhanced, playing an increasingly prominent role in driving economic and social development and laying a good foundation for future development. Bio-medicine, bio-agriculture, bio-manufacturing, bio-energy and other industries began to take shape, and the output value of the national bio-industry has maintained a growth rate of more than 20%. The implementation of protein biological drugs and vaccines, biological breeding and other special projects promoted the development of biotechnology industry. A number of new industries with international standards, such as monoclonal antibody, high-intensity ultrasound focused treatment system and high-end medical diagnostic services are developing rapidly.

4.5 Emerging Industries in Major Economies

4.5.1 Incorporating the Emerging Industries Development into National Strategy

The global industrial development since the first industrial revolution shows that innovation and competition are the main driving forces of industrial evolution. With the faster technological development and higher cost of technological research, it is far from enough to rely on enterprises to promote innovation. The government plays an important role in innovation activities. Emerging industries are characterized by strategic, uncertain, positive externalities and complexity, and are prone to market failure, which needs government intervention. In the initial stage, emerging industries are prone to disadvantages such as insufficient technology R&D capacity, lack of supporting capacity of domestic industrial departments and limited domestic market capacity. If the market mechanism can't give full play to the role of resource allocation, the positive role of government intervention will be highlighted.

In order to promote national technological innovation and the development of emerging industries, so as to occupy a favorable position in the global economic competition, developed economies such as the United States, the European Union and Japan, as well as developing economies such as BRICS, have taken the development of emerging industries to the height of their national strategies. The introduction of a series of programmatic documents has not only made a systematic plan for the development of emerging industries, but also enabled various countries to implement some operational policies to encourage and support the development of emerging industries.

4.5.2 The Key Emerging Industries and New Energy

The development of new technology gave birth to a large number of emerging industries and related industries. Every time a major new technology is introduced into the economy, there is a certain amount of economic growth. If we can accurately capture the possible path and market prospect of the development of new technologies and invest around the revolutionary technologies, once major breakthroughs are made in technological innovation, new technologies and products will not only replace the original technologies and products, but also create new markets and form new industries. Countries have chosen key technologies and emerging industries with great technological potential and market prospects for support, so as to take the lead in the future changes in the global pattern. Sustainable development has become one of the goals pursued by all countries. Therefore, energy conservation and environmental protection, clean energy and biological science have become key industries. The traditional economic development mode characterized by high energy consumption and pollution has posed a severe challenge to the sustainable development of the global economy. It has become the consensus of all countries in the world to transform the economic development mode of the global economy to low energy consumption, low carbon and low pollution. Advances in biological sciences have made it possible for people around the world to effectively prevent disease and improve their life quality.

4.5.3 Human Resources as a Solid Foundation for Innovation

Abundant human resources are the necessary conditions for scientific and technological progress and the development of emerging industries. With the deepening of globalization, the investment in technological research and emerging industries development among the world's major economies is increasing, the international competition of scientific and technological human resources is becoming increasingly fierce. The United States has cultivated a large number of scientific and technological talents through its education system at different levels, and has been spared efforts to import scientific and technological human resources from other countries. In order to retain their own research elite and attract more foreign experts, European countries were also heavily funding the R&D and innovation activities. Since 2008, the European Research Council has provided grants to more than 2,200 outstanding young scientists and engineers through the Young Talent Program.

Emerging economies are also improving their research environment and implementing incentives to prevent brain drain. In 2010, the Russian government set up a special research program for colleges and universities with a total amount of 12 billion rubles (about 400 million dollars) to encourage Russian colleges and universities to carry out joint research and development with foreign scientists, so as to make full use of foreign advanced intellectual resources. In an effort to attract top talent from around the world, South Africa has launched the Chief Scientist Program, in which the chief scientist receives 2.5 million to 3 million rand (370,000–440,000 dollars) a year in government funding for his five-year tenure. India, one of the world's leading sources of scientific and technological manpower, has also introduced measures to reduce the brain drain, including setting up the Tata Technology Innovation Prize and allowing scientists to receive commissions based on their contributions in R&D projects. With the rapid development of emerging industries and the changes in the supply and demand of scientific and technological human resources, the major economies in the world are adjusting their policies trying to constantly improve the cultivation and utilization of scientific and technological human resources.

4.5.4 Effective Policy and the Development Law of Emerging Industries

In the long-term development process, developed economies have formed an institutional environment conducive to innovation, these countries still seek whether there are factors that hinder the innovation efficiency and put forward systematic solutions to existing problems so as to attract more resources to gather in the innovation. In the Europe 2020 Strategy Period Plan, the EU pointed out that in the post-crisis era, Europe must break the "business as usual" development model, take innovation as the overriding policy goal and reform the patent system to make intellectual property rights become a source of competitive advantage for enterprises in developed economies. The United States has also set up a number of innovation centers to provide opportunities for scientists and entrepreneurs to work together on innovation and speed up the transformation of technological achievements.

According to the characteristics and laws of the development of emerging industries, the government gave incentives to the development of emerging industries mainly from the two aspects of supply and demand. In terms of supply incentives, major economies reduced the costs and risks of technological innovation and transformation by providing subsidies for technological R&D and to technology and product suppliers and unifying technical standards meanwhile. Continuously expanding the scale of emerging industries, developed economies also actively promoted the formulation and application of international standards in relevant fields. In November 2011, the United States, the European Union and Japan agreed to work together to advance international standards for electric vehicles within the framework of the UN/WP29 forum for Harmonization of Vehicle Law. The formulation and implementation of industrial standards for emerging industries not only promoted the industrialization of new products, but also enabled developed economies to stand at the high end of the value chain through the formulation of industrial standards. In terms of demand incentives, major economies have also strengthened public procurement, demonstrated product application and provided consumer subsidies, so as to stimulate market demand for new products and provide more market space for emerging industries. In order to start the electric vehicle market, the US Department of energy provided us $99.89 million to implement the "EV project" demonstration and promotion project of electric vehicles and charging facilities, deployed 11,210 charging stations in 11 major cities and installed chargers for car owners free of charge.

4.6 Impacts of Emerging Industries

4.6.1 Emerging Industries and Industrial Development Pattern

Technology is developing at an unprecedented speed, which has greatly promoted the rapid growth of emerging industries. Driven by technological progress, a number of emerging industrial sectors are emerging on the basis of high and new technologies such as energy conservation and environmental protection technology, clean energy technology, biotechnology, information technology, new material technology and advanced manufacturing technology. At present, major economies are focusing on the development of new energy industry clusters and biotechnology industry clusters are in the initial stage. With the rapid development of emerging industries, a new pattern of industrial development with new energy and environmental protection industry, biotechnology industry and related high-tech industry as new economic growth points will be formed in the near future.

The development of emerging industries will also have a certain impact on traditional industries. For example, the increasing popularity of electric vehicles will challenge the industry chain that has long been formed in the traditional automobile manufacturing field with engines, gearboxes and on-board electronic equipment as the core. More and more suppliers will join the emerging industry chain with power batteries, drive motors and electronic control fields. However, the development of emerging industries does not mean the demise of traditional industries. Emerging industries have strong penetration. With the continuous maturity of clean energy and low-carbon environmental protection industries, these technologies will spread to other sectors, and their penetration and driving role in traditional industries will be increasingly enhanced. If the traditional industry department using new technology to support to modify work processes and organizational structure and to carry out a wide range of process innovation, product innovation and business model innovation, then it can improve the efficiency of elements combination of production, transform the quality of products and services. Eventually, traditional and emerging industries will work together to achieve sustainable economic development based on low carbon and environmental protection.

4.6.2 New Industrialization and the "Core-Periphery" Position

In accordance with the "core-periphery" Waller's theory, the developed countries in the world economy took control of the production, capital, technology and trade, thus in a dominant position, become the core area of the world economy. Developing countries were attached to the edge of the area of developed countries in order to obtain production, capital, technology and trade from the core areas. But the continued growth and rise of the non-western world today brings a different interpretation to Wallerstein's theory. Western developed countries have too much excess capital and need to look for investment opportunities, whose manufacturing industries need to be transferred due to high labor and welfare costs. The new round of industrialization in developing countries provides these opportunities and conditions.

First, developing countries like Brazil have achieved political stability after decades of national construction since they got rid of colonial rule and gained independence in the 1950s and 1960s, and can provide basic conditions for economic development, such as compulsory education and transportation. Qualified and cheap labor can only be found in newly industrialized countries, which is an important reason for the transfer of manufacturing and production links to developing countries. Second, the industrialization of developing countries has brought huge investment opportunities and broad market demand. In particular, industrialization has brought huge capital demand, production (raw materials, energy, equipment, technology) demand, infrastructure demand and trade demand. Although the new round industrialization of developing countries is mainly based on manufacturing industry, its advantages in manpower, large-scale production and consumption demand, industrial demand and investment demand have made developing countries become the core demand area of world production and the core area of world economic growth.

Looking towards the global economy, the industrialization of the world development, including the manufacturing industrialization, is far from being completed. At present, apart from OECD countries, non-OECD economies and a few countries in transition such as Russia have partially realized industrialization, there are more than 130 developing countries in the world, whose land area and population account for more than 70% of the world's total. They urgently need to realize economic growth and get rid of poverty through industrialization. According to estimates by the United Nations Conference on Trade and Development, the financing gap for industrialization in developing countries is as high as 1.9 trillion dollars to 3.1 trillion dollars annually, including 1.6 trillion dollars to 2.5 trillion dollars in infrastructure investment. In response to this strong demand for industrialization, investment flows to developing countries have grown rapidly over the past decade, with private capital investment alone increasing more than tripled, reaching 327.7 billion dollars in 2013. In recent years, with the large-scale overseas investment of China's "One Belt and One Road" construction project and the active South-South cooperation carried out by India and other emerging powers, the investment flowing to developing countries was increasing rapidly. It well confirmed that new round of industrialization led by a group of newly industrialized countries such as China is continuously spreading from industrialized countries to developing countries. It is no exaggeration to say that the new round of industrialization rising in developing countries since the new century has a wide range, large scale and deep degree, which is far beyond the period of primary industrialization and late industrialization. It is really worldwide group industrialization and will also bring about great global changes in a more fundamental sense.

4.6.3 The Return of Nationalistic Governance Paradigm

Since the 1980s, economic globalization and liberal economic policy have become the trend of international and domestic governance, but the return of nationalistic governance has become a prominent feature of the global change that we are experiencing. First, the industrialization of developing countries was mainly driven by national strategy, which was different from the industrialization driven by liberal capitalism. The industrialization of developing countries has obvious characteristics of government-driven. Some countries even have emerging industrialization driven by national capital and government departments.

The second is the return of global governance to national governance. In many key areas of global governance, the capital needs of global governance are enormous, and the biggest beneficiaries of governance investment remain individual sovereign states. For example, global climate governance requires a huge amount of capital to promote the transformation of national economy to low-carbon economy. This huge interest chain of who invests and who benefits ultimately promotes the generation of national governance plan of Paris Agreement. The Millennium Development Goals, the Post-2015 Development Goals and the 2030 Agenda for Sustainable Development take development, poverty eradication and low carbon emission reduction as their core tasks, and take into account national development goals and the common interests of the world. However, it is clear that individual sovereign states are the main and most direct beneficiaries. Therefore, both the Millennium Development Goals and the 2030 Agenda for Sustainable Development place national programs are at the core of global governance. Moreover, the gap between the funding required for global climate governance and the global development agenda and the need for aid, even the smallest amount requested, has never been met. Therefore, the return of global governance to national governance scheme is inevitable.

Chapter 5
Industrial Transformation and Reconstruction

This chapter introduces the theory of industrial transformation from the perspective of industrial chain in the context of globalization. Industrial chain is the micro composition of industrial structure, and industrial structure is the superposition of many industrial chains. The change and adjustment of industrial chain will promote the adjustment and transformation of industrial structure. With the continuous development of technology, industry chain competition appears. The industrial chain competition not only expands the industrial chain division of labor, but also achieves the maximum governance rent. However, as COVID-19 broke out, production and logistics have stalled and regional blockades have changed the original pattern of globalization. COVID-19 has hit global economic and trade activities and disrupted the original global industrial chain. Therefore, in the post-pandemic era, we need to restructure the global industrial chain and realize the aggregation and new development of the industrial chain driven by technology, market, government and cost. Finally, the thoughts and paths of industrial chain reconstruction are more detailed through the value chain analysis of capital-intensive industry, the automobile industry.

5.1 A New Theoretical Perspective of Industrial Transformation

The original theory of industrial structure is based on static or comparative static analysis, without considering the resource transfer between industries, which lacks certain dynamic characteristics. With the deepening of globalization, the change of labor division and the transfer of resources between industries, scholars explored the impact of international labor division on industrial structure from a dynamic perspective. From the perspective of industrial chain, this chapter brings technological innovation and international labor division into the study of industrial transformation so as to help readers more clearly understand the logical relationship between technological change and industrial transformation.

Y. Jiang and J. Gu, *Technology and Industrial Transformation of China*,
https://doi.org/10.1007/978-981-19-7458-8_5

5.1.1 Industrial Chain

The industrial chain is a concept put forward by Chinese scholars. It was born in the 1980s when Chinese scholars studied the industrialization of agriculture. Later, scholars extended it to industry, service and other fields. However, the idea of industrial chain could be traced back to the study of the division of labor by classical mainstream economists in the middle and late eighteenth century. In the search of literature review in this book, Jacobs, an American scholar, first used "industrial chain". In *Cities and the Wealth of Nations*, Jacobs discussed the multiplier effect of industrial chain on the economic growth of a region. In China, there were many research literatures related to industrial chain, but some mistakes appeared in the use of industrial chain and the understanding of concept connotation. This section first summarizes the views of some domestic scholars on industrial chain, and then analyzes the conceptual level of industrial chain.

5.1.1.1 Domestic Scholars' Understanding of the Industrial Chain

From the product point of view, the industrial chain equals to the value chain. Influenced by Porter's concept of value chain, some scholars in China expanded or combined the definition of value chain on the basis of the concept in the study of industrial chain or in the papers with "industrial chain" as the theme. Industrial chain refers to the industrial value chain, and the value chain refers to the process of industrial value creation can be decomposed into a series of interrelated economic activities (Xu 2007; Zhao 2008). Industry is a collection of products of the same nature. It is no logically right in defining the industrial chain from the perspective of products. However, in the condition of labor division, such definition wrongly equates the value chain to the industrial chain.

From the perspective of connection, industry chain is the expression of industrial relationship and inter-enterprise relationship. Industrial chain is a set of industries with internal links (Lu et al. 2004). For a region, the integrity of the industrial chain structure affects the industrial supporting capacity and the development of the whole industry (Wang and Li 2006). From the perspective of inter-enterprise relations, the industrial chain is regarded as the strategic alliance relationship between enterprises and the chain of strategic alliance relationship, or the vertical relationship (Yu and Guan 2006) and the horizontal collaboration relationship. From the perspective of leading enterprises, industrial chain refers to the enterprise symbiotic value system composed of multiple enterprises.

From the perspective of organization, industry chain is the organizational form of labor division. Industrial chain is an industrial organization form based on the labor division economy, which emphasizes the labor division and cooperation between related industries or enterprises. This labor division and cooperation is represented by the supply-demand relationship between enterprises (Liu 2005). As a trans-regional organization of production cooperation, industrial chain is connected with enterprises through contractual relations (Li and Wei 2007). Industrial chain is a functional network chain based on knowledge division and collaboration, which is an industrial optimization combination formed to improve the efficiency and competitiveness of the economy (Liu and Weng 2007).

5.1.1.2 Logical and Economic Concepts of Industrial Chain

Domestic scholars' understanding of industrial chain is mainly carried out from three perspectives, product, subject (industry) relationship and activity organization, which has not formed a relatively complete conceptual structure. The following part of this book will analyze the hierarchical structure of economic activities, so as to better reveal the conceptual structure of the industrial chain. Production in an economic system is mainly carried out by a number of the most basic activities through the way of interlinkage, which is called the organization of economic activities. The productive sector or the consumption sector has different forms of economic activity organization. No matter which form, in essence, it reflects the difference of behavior nature and geographical difference of activity scope of economic actors in the process of trading. The difference of behavior nature shows that the subject's behavior is subject to different constraint rules and incentive mechanisms. The geographical difference in the scope of activities shows that the connection rules between activities are different.

There are three main levels of economic activity. The first is the activity layer, which is the basic layer for the production of any product or service. It is composed of inter-linked production activities of various departments within the enterprise, namely the "value chain" said by Porter. Each activity needs corresponding production factors to match it. Internal value activities of enterprises can also be independent of enterprises through certain rules (outsourcing, contract manufacturing), such as design, production, logistics, warehousing and other links are increasingly socialized into an industry. In this case, the concept of industrial chain and value chain is easy to confuse. The second layer is the enterprise layer. The result of the activity level is the product output at the enterprise level, which is either used as an intermediate input of other products or as a final consumer product. No matter at which stage, the collection of products constitutes the corresponding industrial layer. The third is the industrial level, which focuses on the collection of series products involved in the production of a product.

The logical definition of industrial chain is based on the condition of labor-division economy, where economic organization has different levels of performance. Industry chain is one that focuses on the industrial level. Conceptually, the industrial chain refers to a complete chain formed according to production logic in the production process of a final product, from the initial mineral resources or raw materials until the final product reaches consumers. Starting from the basic concept, the industrial chain is the objective expression of industrial labor division. For economic research, what is more important is the action significance that this division of labor brings to the subject.

5.1.2 *Industry Chain as the Microcosmic Composition of Industrial Structure*

The industrial chain may be oriented by enterprises that produce final products or intermediate products. Of course, it can also be composed of enterprises that produce or provide products and services. All industrial chains are actually the basic links between technological and economic supply and demand of all enterprises in the process of production and services. If these links are decomposed vertically or horizontally, they are really what we call industries. The industry here refers to the collection of similar enterprises (same products, same technology, same technology, same service object, etc.). These industrial clusters (such as generators, automobiles, clothing, etc.) constitute the industrial structure of a country or region and the input-output relationship from primary products, intermediate products to final products.

5.1.3 *Industrial Chain Competition*

Competition is the key driving force of economic development and the structural attribute of market economy. The emergence of industry chain competition is based on two basic changes in the competitive environment of enterprises. The first is a shift in the content of the competition. The content of enterprise competition changes from "product supply" to "service supply". In the era of large-scale, mass-produced industrial economy, enterprises are concerned about the availability of products, while consumers are concerned about the availability of services. The second is a "shift in the level of competition". The competition level of enterprises has changed from individual enterprise competition to industry chain competition. Industry chain competition is a more advanced form of integrated competition development and a dynamic competition behavior, which aims to expand the division of labor of industry chain and strive to create new modes of production, new productivity and new demand.

5.1.3.1 Industrial Chain Competition and the Labor Division

As a more advanced form after the competition evolution, industry chain competition can be distinguished according to the degree of labor division. The first is market competition and enterprise competition. Market competition means that there is no monopoly power in the market (Stigler 1942). It refers to the competition between enterprises for similar products in the established market structure and the result is that the product price tends to the natural price. Industrial competition is a concept first used by Carines. It refers to the flow of resources or capital of the same property between industries of different property, resulting in the tendency of factor rates return to be equal. Carines' understanding of industrial competition is equivalent to what we now call perfect competition. Market competition is the short-term behavior state of enterprises in a given market, while industrial competition is the long-term behavior state of enterprises in inter-industry competition. Second, there is Ricardian competition and integration competition. Ricardian competition refers to the competition between firms in a similar commodity market for a given market share. In this competitive situation, some enterprises adopt predatory pricing strategy in order to exclude competitors. Integrated competition first put forward by Clark, he thought under the Ricardian competition, the competitive behavior between enterprises with similar products, but in these enterprises integrate competition stop attacking each other and choose a different form of system, collective action and the industrial chain of upstream or downstream of the enterprise group.

At present, the competition among enterprises is changing from Ricardian competition and integration competition to industry chain competition. This transition is based on the change of competition content, from the original product supply to service supply change. In the industrial chain competition, the industrial organization mode driven by production technology will connect or give way to the industrial chain structure and production organization mode based on consumer demand, which is an important development strategy for enterprises to control the power in the era dominated by diversified and personalized demand. Industrial chain competition, as a dynamic competitive behavior, can form the overall competitiveness of the industrial chain by choosing appropriate industrial chain structure and transaction structure, constantly expand the division of labor of the industrial chain, lead enterprises in the industrial chain to continuously seek and achieve new profit growth.

5.1.3.2 Industrial Chain Competition and Governance Rent

Bullock (1902) called the principle of realizing mass production "the law of organizational economy", that is, the reduction of marginal cost realized by enterprises through the change of production organization scale. Marshall (1913), who extended the Bullock economic law to make it consist of a single enterprise scale expansion to assemble in industrial spatial scale, thought organization economic law besides enterprise scale can be expanded, similar industry spatial agglomeration scale would also promote the organizational improvement, bring the change of the enterprise on the factors of efficiency, namely the "law of increasing returns. Different from Bullock's "law of organizational Economy", Marshall's aim is to improve the efficiency of factor use. This kind of efficiency came from the external economy generated in the development of specific industry in location agglomeration.

Based on the above analysis, Ricardian competition and integrated competition strive to obtain a larger market share under a given industrial labor division framework and market scale. Industry chain competition is to strive to create new productivity, new demand and expand market scale. The behavior subjects are composed of heterogeneous enterprises in each node industry, and each subject has different competitive behaviors. In order to adjust, control and protect these different behaviors, innovation rules are needed. Therefore, industry chain competition is based on the law of organizational economy, through changing the composition and governance of industry chain, so as to expand the market scale, reduce transaction costs, and create and occupy innovation rents.

5.2 Global Industrial Chain Development Before the COVID-19

In the past 30 years, globalization has two characteristics. One is that western developed countries transferred the processing and manufacturing of finished and semi-finished products, components and raw materials to developing countries, and used capital and technology to obtain high profits, forming industrial division of labor among countries. Another is that multinational companies outsourced the processing of some finished and semi-finished products in the industrial chain to global enterprises with good processing quality, high efficiency and low cost according to the theory of horizontal labor division, forming a global industrial chain division of labor. As a result, dozens of countries, hundreds of cities and thousands of enterprises jointly participated in production, while multinational companies grasped the R&D and design in the upstream of the value chain, the sales settlement and after-sales service in the downstream and obtained high profits by controlling the productive service process. Such a horizontal labor division industry chain system promoted the optimal allocation of resources in a market-oriented way and realized the goal of maximizing profits for multinational corporations.

The theory of horizontal labor division was first systematically elaborated by Thomas Friedman, an American economist, in his book *The World is Flat*. From the perspective of economics, it is quite reasonable for all kinds of enterprises in the industrial chain to conduct horizontal labor division according to their own advantages regardless of countries and regions, which is conducive to the reduction of product costs. However, as a matter of fact, the earth is round and the world is not always flat. Too many links in the industrial chain and long transportation distance will inevitably lead to high logistics costs and long transportation time, thus increasing the risk of the rupture of the whole industrial chain. Due to the simple pursuit of profit in the horizontal labor division, each link of the chain has used various information technologies to run the information flow, logistics and capital flow of the industrial chain to the extreme. All chain members are complacent about the high efficiency of the horizontal labor division global industrial chain in the period of stable operation, and rarely consider the huge risks inherent in such a well-run chain in the event of a global event. A local event will not completely affect the balance of the original industrial chain. If a regional production enterprise stagnates, the industrial chain will automatically seek replacement in other parts of the world.

Combining the power of market, technology, cost and government, the development of global industrial chain before the COVID-19 mainly has the following three characteristics.

5.2.1 *The Shift of Labor-Intensive Industries*

Labor-intensive industries are sensitive to cost changes, but have relatively low requirements on technological level, capital equipment, industrial supporting facilities and infrastructure. Therefore, with the development of economy, labor force in this industry shows the characteristics of transferring from high-income countries to low-income countries. The shift has expanded from traditional industries such as textiles and clothing to labor-intensive processes that can be separated and outsourced in electronics, chemicals, transportation equipment and machinery. Take the textile and garment industry as an example, since the industrial Revolution, this industry has experienced five times of transfer from Britain to the United States, from the United States to Japan, from Japan to the "Four Asian tigers", from the "Four Asian tigers" to the "Tiger Cub Economics" and China, and finally from China to other Southeast Asian, South Asian and African countries.

Low-priced labor helped China integrate into the global labor division. Processing trade is one of the important ways for China to participate in the division of global industrial chain, which combines the advanced production technology and management experience of multinational companies with the advantages of local labor and natural resource endowment, thus achieving a win-win situation. In 1990, China's population between the ages of 15 and 64 accounted for 23.3% of the world's total labor force. Its abundant labor force, land and other resources provided a solid foundation for the development of processing trade. From 1995 to 2007, processing trade accounted for more than 45% of China's total imports and exports, higher than general trade. Labor intensive industries and processing trade drove China's economic growth, and also the development of capital and technology-intensive industries. At the same time, with the integration of China, the global industrial chain is gradually shifting from the "Asia-Pacific-Europe and Africa" bipolar model to the "three pillars" pattern of North America, Europe and Asia.

In recent years, labor costs in China have risen rapidly, driving labor-intensive industries to Southeast Asia, South Africa and Africa where labor costs are lower. According to a survey by the Japan International Trade Promotion Agency, the monthly basic salary of the average worker in Shanghai in 2014 was 1.15–6.97 times that of the corresponding cities in Southeast and South Asian countries, with the biggest gap with Myanmar and Vietnam, and even more than 10 times that of some African countries. In the textile and garment industry, for example, although China is still the leading exporter of final consumer goods and intermediate inputs, there are signs of a shift to Southeast Asia, South Asia and Africa. From 2007 to 2018, the share of apparel exports from other Southeast and South Asian countries in the world rose from 9.7 to 13.5%, indicating a significant increase in growth and may be further accelerated.

5.2.2 The "Double-Edged Sword" Effect on Industrial Chain Development

On the one hand, standardization, modularization and digitalization have greatly enhanced the degree of "diffusion" of complex technologies, provided opportunities for developing economies to integrate into globalization and promoted the establishment of a global division of labor system. The development of standardization, modularization and digitization technologies has drastically reduced the investment in R&D, Learning by Doing and other complementary skills required for production. Digital technologies facilitate the integration of developing countries and small and medium-sized enterprises into global value chains (GVCS), including reducing the cost of information transfer and cross-border transactions, increasing the availability of intermediate goods, services and technologies, and focusing on local production processes. From a macro perspective, the trade between developed economies and emerging economies has been strengthened in the past 20 years, especially the trade between European, North American and Asian industrial chains. From a micro perspective, China is increasingly important as a supplier to multinationals for companies from the developing world. Take Apple Inc. as an example, 40 of its top 200 suppliers were from China, and 47.5% of the 807 factories in the world were in the Chinese mainland in 2019.

On the other hand, emerging technologies constantly change the production mode of traditional industries, increase the technical barriers to economic catch-up and even replace the factor advantages of developing countries. Skill-biased technological progress leads to the substitution of low-skilled labor force, which weakens the comparative advantage of developing countries in traditional labor-intensive industries. In particular, companies that use industrial robots in manufacturing are less likely to outsource their manufacturing activities to other countries. Within industry, the flow of foreign direct investment in electronics and automotive products with high automation intensity is significantly lower than that in textiles, clothing and leather products with low automation intensity. In addition, in the era of digital economy built by the Internet, data and artificial intelligence, mass customization, dynamic industrial chain, intelligent production and service have become new features of modern industry, which is difficult to be directly imitated by low-income countries engaged in mass production with low-priced labor.

5.2.3 Some Manufacturing Being Brought Back

The formation and development of the existing global industrial chain depends on a positive and stable external environment. After the global financial crisis in 2008, the world economy entered a period of deep adjustment, with populism and trade protectionism rising and anti-globalization surging. The global industrial chain shows signs of regionalization and even localization.

In Asia, the share of intraregional trade in The Asian industrial chain increased from 40.3 to 46% from 2000 to 2017. In forward and backward industry chain activities, intra-regional trade in Asia increased by 3.5 and 7.4% points respectively. In forward and backward complex industrial chain activities, the share of intra-regional trade in Asia increased by 5.4 and 6.6% points respectively. An important factor contributing to this change is the integration of a large number of low-and middle-income Asian economies into Asian production networks. The intra-regional trade of computer, electronic and optical products, chemical and pharmaceutical industries, coke and refined petroleum products accounts for a relatively high proportion, while the extra-regional trade of mining, agriculture, forestry, animal husbandry and fishery and other transportation equipment industries accounts for a relatively high proportion. In the future, driven by cooperation agreements such as the Belt and Road Initiative and the Regional Comprehensive Economic Partnership, the degree of regional economic integration in Asia is expected to be further enhanced.

In North America, the demand for reshoring is increasing in labor-intensive industries and industries mainly export to US. For one thing, after the financial crisis, developed countries began to reflect on the problem of "hollow phenomenon of manufacturing" and put forward the slogan of "re-industrialization" in succession in order to realize the return and revitalization of manufacturing industry. For another thing, as China rapidly entered the high-tech industry and began to threaten the global dominance of the United States, the United States identified China as the main competitor and constantly increased the attack on Chinese industry, to avoid and prevent China from catching up in the high-tech field. In the influence of Sino-US trade frictions and US-Canada-Mexico Agreement, labor-intensive industries and industries that mainly export to the US have accelerated their withdrawal from China to North America. From January to July 2019, US imports were basically flat compared with the same period in 2018, but the structure of its trading partners underwent significant adjustment. Imports from China decreased by us $36.6 billion, while imports from Mexico increased by 12.5 billion dollars.

5.3 The Impact of COVID-19 on the Industrial Chain

With the development and integration of technology, economy and society, the linkages of economic and trade in various countries are becoming closer with the promotion of multinational corporations which make full use of resources like technology, capital, labor and natural resources on a global scale. However, the outbreak of the COVID-19 in 2019 cast a layer of haze on the global economy, which was in recession. The shutdown of factories, as well as the rupture of people flow, logistics and capital flow, forced the operation of multinational corporations to stagnate, and the global economy and trade fell into a quagmire, bringing out the trade protectionism and unilateralism. The decline of global economy and trade nourished the trend of anti-globalization. The isolation and disruption of countries and regions has had a huge impact on the global industrial chain and supply chain. The suspension of production has led to the loss of original product orders of enterprises in the chain and even those enterprises that are not affected by the pandemic have lost their original parts supply due to logistics disruption.

5.3.1 Trade Control and Further Deepening of Globalization

With the spread of COVID-19 around the world, major economies in the world have been hit hardest by the outbreak, prompting countries to adopt trade controls and embargoes, which not only severely undermined the free trade system in the WTO framework, but also interfered with international trade of medical and pandemic prevention. In the long run, it will undermine the existing trade and investment system, discourage countries' confidence in trade and investment liberalization, and hinder the in-depth development of economic globalization, which is embodied in two aspects. First, the embargo of the major economies resulted in the stagnation of the flow of elements of human capital. In response to the outbreak of COVID-19, major economies have implemented policies and measures to ban flights and embargoes. On 12 March 2020, following the declaration of COVID-19 as a global pandemic by the World Health Organization (WHO), additional embargoes were imposed by major economies. The U.S. Government declared a state of emergency in the United States on March 13, 2020 and suspended regular visa operations at its embassies worldwide on March 20. Since March 17, Germany has implemented the EU's 30-day ban on entry of citizens from non-EU countries, which applied to all aircraft and ships departing from outside the EU. Russia banned all foreign nationals and stateless citizens from entering Russia. The South Korean government announced that passengers from five countries, France, Germany, Spain, the United Kingdom and the Netherlands, would have to undergo further entry quarantine procedures. Embargoes in these economies have severely disrupted the international shipping system and hampered the flow of human capital.

Second, the commodity trade control measures of major economies have greatly restricted the global flow of materials and funds. As COVID-19 spreads around the world, countries have introduced trade control measures to limit trade flows and severely undermine the free trade order. The global shutdown and the suspension of trade and investment have led to the rupture of the global supply chain and the serious shortage of medical materials in various countries, threatening the national public health security. At the same time, countries have adopted stricter standards and control measures for entry-exit inspection and quarantine of commodity. By May 25, 2020, a total of 35 countries had notified the WTO of their control measures on trade in commodity. Among them, the United States, India and other countries have strengthened supervision over imports of Chinese products, taking many inspection measures, such as import screening, inspection, sampling and import early warning. Italy and Germany required Chinese exporters to provide documents proving that their products are free of novel Coronavirus contamination, etc. More burdensome entry and exit measures hurt international production, trade and investment.

5.3.2 Global Economic and Trade Under the COVID-19 Pandemic

Production stopovers and lockdown measures taken by various countries to prevent and control the pandemic have inevitably impacted the global economy, trade and investment, with production capacity, demand for trade in goods and services shrinking significantly. Shipping embargoes and trade controls have seriously threatened international transport, international travel and trade in goods, disrupted the flow of people, logistics and capital, and dealt a major blow to cross-border trade. On May 20, 2020, the Barometer of trade in goods released by the World Trade Organization (WTO) showed that the volume of world trade in goods dropped sharply in the first half of 2020. The index was only 87.6, far below the benchmark value of 100, which was the lowest record since the index was launched in July 2016. According to WTO statistics, global trade in goods fell by 5.3% in 2020. According to OECD statistics, the exports of major countries and regions in the first quarter of 2020 fell sharply from the same period last year (Table 5.1).

The entry and exit of people and goods were strictly controlled. Multinational corporations and their overseas subsidiaries delayed capital expenditure and postponed or shelved existing investment plans, which also had a non-negligible impact on global investment. Global venture capital funding has fallen by 20% since the COVID-19 outbreak, with an even more pronounced decline in China where funding fell by more than 50% in January to February 2020 relative to the rest of the world. Shutdowns have also led to large numbers of small- and medium-sized enterprises going bankrupt and unemployment rising. Historical experience shows that

Table 5.1 Exports and growth rates of major countries in the first quarter of 2019 and 2020 (100 million dollars)

	France	Germany	Italy	Spain	UK	China	USA	India
2019	1452.7	3819.9	1302.4	843.8	1155.9	5511.7	4085.1	860.0
2020	1298.9	3586.4	1239.8	783.5	1024.6	4775.9	3956.8	750.3
Growth rate	−10.6%	−6.1%	−4.8%	−7.2%	−11.4%	−13.4%	−3.1%	−12.8%

Source OECD Statistic database

the process of economic globalization is closely related to the development of the global economy and trade. When the global economy as a whole improves, the process of globalization will accelerate, and when the global economy is sluggish, the process of globalization will slow down. The continued spread of COVID-19 has led to a global economic and trade recession, which is bound to further intensify the anti-globalization trend.

5.3.3 Disrupted Global Value Chains

In recent years, the process of globalization is accelerating and its essential feature is the formation and development of global value chain. Multinational companies closely integrate the production chain of all countries in the world, deploy the R&D, production and sales of products to all countries with the lowest production and transaction costs, and form a global network of production, trade and finance. Compared with the SARS outbreak in 2003, the global industry is now more integrated, with global supply chains accounting for 28% of trade and 46.7% of global industrial production. The international labor division with comparative advantage promotes the rapid development of global trade and investment. The deepening international labor division not only reduces production cost and improves production efficiency, but also increases the vulnerability of global production.

The global spread of COVID-19 has hit the core sectors of production hard, especially in developed regions with large population, trade and capital flows. The world's three largest production networks, East Asia, Europe and North America have been hit hardest. Currently, confirmed COVID-19 cases are mainly concentrated in countries and regions with high GVCS participation index, such as the US and Europe, while there are few confirmed cases in countries with low GVCS engagement index, such as Cambodia and Laos. Because of the huge impact of the COVID-19, single-source suppliers or regions specializing in specific products expose unexpected and far-reaching vulnerabilities, making GVCS to break down and collapse.

In 2008, the global economy was in a prolonged period of sluggish recovery due to the financial crisis and the trend of anti-globalization appeared. Since Trump took office in 2017, he has pursued "unilateralism" and trade protectionism around the world, which has impacted the foundation of global economic and trade cooperation. During the pandemic, major economies have been reflecting on the cost of global value chain integration and international labor division. Based on the goal of national security, some countries and regions emphasized inward development and independent development. Some multinational enterprises also shortened the links of the international supply chain based on security factors to prevent the supply interruption caused by the long value chain.

5.4 Reconstruction of the Global Industrial Chain in the Post-pandemic Era

The relationship between COVID-19 and globalization is the basis for assessing the future direction of global industrial chain restructuring. The COVID-19 pandemic has led to a simultaneous decline in global supply and demand, value chains broken and the divergence and confrontation of ideologies and values among countries, which is changing the foundation of globalization characterized by the transnational allocation of factor resources. For the trend of globalization after the pandemic a comprehensive literature review found the following representative views. One is limited globalization. Countries have regained more economic sovereignty, and the economy has returned from the "hyper-globalization" characterized by the rapid flow and optimal allocation of factors of production around the world to the "limited globalization" characterized by the flow of goods and capital before the 1980s. The second is re-globalization. On the basis of the economic structure "center-periphery", economic structure and represented by China, due to the willingness and ability of the global governance, the expansion of domestic market has gradually replaced the western powers as the new driving force of globalization. The "Dual pattern" of globalization composed of developed and developing countries have evolved into "Ternary pattern" composed of developed countries, emerging countries and peripheral countries. The third is slow globalization. As the global flow friction of goods and factors of production intensifies, the growth rate of global investment slows down and the conflicts of interests between multinational companies and their parent companies increase, the trend of globalization slows down, the degree of regional economic integration deepens, and the global value chain shrinks inward. The fourth is digital globalization. The digital economy and artificial intelligence will replace trade in goods and finance as the main drivers of globalization, represented by increased connectivity online and a slowing of the flow of factors offline.

The "hyper-globalization" dominated by developed countries and characterized by the rapid flow of factors of production across the globe has receded. Globalization will continue to be supported by the development of digital economy and artificial intelligence, the enhanced capacity and willingness of emerging economies to participate in global governance, the objective needs of developing countries in the process of industrialization and integration into the global economic labor division. In the future, it will be a globalization in which various forces participate in and balance each other. The global industrial chain will also seek a new steady state under the balance of technology, market, cost, competition, government and other factors. On the one hand, market, technology and cost factors continue to drive the development of the global industrial chain. From the perspective of market, the growing domestic market of developing economies represented by China will attract transnational corporations to form new regional agglomeration mode centering on terminal demand. In terms of technology and cost, the global industrial chain will become more knowledgeable, digitized and capitalized as the global factor endowment pattern changes. On the other hand, from the perspective of government and policy, the influence of safety orientation in the global industrial layout has increased significantly. For one thing, the impact of COVID-19 accelerated the trend of regionalization and localization of the global industrial chain that had emerged after the financial crisis. For another thing, the outbreak of COVID-19 prompted multinational companies to implement diversified and concentrated strategies in the layout of the industrial chain.

5.4.1 Regional Aggregation of Industrial Chain and Terminal Market Demand

The geographical pattern change of global demand is one of the important forces reshaping the global industrial chain. According to the McKinsey report, before the financial crisis, the recipient economies were globalized consumers. More than 74% of global consumption in 1999–2007 came from developed economies. After the financial crisis, the share of developing economies in global consumption rose rapidly. In 2017, 38% of global consumption came from developing economies, up to 12% points from 2007. Among them, China, developing Asia excluding China and developing America contributed 6, 2 and 2% points respectively. According to McKinsey's prediction, China, developing Asia excluding China and developing Europe will continue to increase their share of global consumption by the same percentage points by 2030, which is bound to attract more multinational companies to form a new regional cluster model around terminal demand.

5.4.2 Technology- and Cost-Driven Industrial Chains

With the change of the global factor endowment pattern, the global industrial chain will show the trend of knowledge, digitalization and capitalization in the medium and long term. Knowledge and intangible assets are becoming increasingly important to GVCS. From 2000 to 2016, R&D and capitalization of intangible assets in GVCS increased from 5.4 to 13.1% of total revenue. Knowledge intensity of all types value chain activities increased to varying degrees, among which pharmaceutical and medical equipment, mechanical equipment, computer and electronics production ratio were the highest (80, 36.4 and 25.4%, respectively). Accordingly, value chain creation moves to upstream activities such as R&D and design, and downstream activities such as revenue and after-sales, while the value share of manufacturing declines.

In addition, automation and artificial intelligence technologies and a prolonged low interest rate environment will make manufacturing increasingly technological and capital intensive. From a technological perspective, emerging technologies such as artificial intelligence and 3D printing are replacing low-skilled labor and changing the way of industrial production. From the perspective of policy environment, the low interest rate policy generally implemented by major developed economies reduces the price of capital relative to labor, and promotes the digitalization and capitalization of industrial chain. In the future, countries with more skilled labor, stronger innovation capabilities and better intellectual property rights protection will play a more important role in the new global industrial chain.

5.4.3 Diversification and Agglomeration of Industrial Chain

In order to ensure the toughness and flexibility of the supply chain and strengthen the anti-risk ability of the industrial chain, multinational companies will implement the strategy of diversification and agglomeration in the layout of the industrial chain. Diversification spreads risk and reduces dependence on a single economy or a handful of foreign suppliers. Agglomeration can reduce transportation cost, shorten logistics time, improve logistics scheduling efficiency and avoid the impact of various natural disasters and pandemic disasters to the greatest extent through industrial chain cluster.

The diversified layout of industrial chain is mainly reflected in the service industry and the complex value chain manufacturing industry. Take Japanese auto industry as an example, its components are highly dependent on China. During the outbreak of the pandemic in China, due to the shutdown of Chinese factories and the interruption of logistics, Japanese enterprises were forced to stop production due to the shortage of components. On March 5, 2020, Japanese Prime Minister Shinzo Abe said that Japan must consider moving its production base for highly dependent and value-added products back to Japan. If this is not possible, they will try not to rely too much on a single country, part of the production of components to Southeast Asian countries to achieve diversification of production base. Diversification increases the opportunities for developing country enterprises to participate in the global industrial chain, but in terms of human resources and infrastructure, China, South Korea, Mexico, some Southeast Asian countries and a few Eastern European countries are the main candidates for industrial chain layout. After the pandemic, developed countries will accelerate the improvement of domestic or regional industrial chains for safety reasons, and may even try to build a "De-Sinicize" global industrial chain. However, as an important place for the development of new technologies, China has a strong industrial supporting capacity and a huge market capacity, which can play an important role in the new global industrial chain in the long run.

Industrial agglomeration mainly focuses on key industries such as national strategic emerging manufacturing industry or national pillar industry. To ensure economic security and improve the resilience of these industries, economies tend to build clusters of the whole industrial chain, from R&D and design to logistics and finance. At the same time, the industrial chain cluster with a large market scale can enhance the voice of the economy in the global industrial ecology, and effectively avoid the possible business risks caused by the "bottleneck" of key technologies with the market scale.

5.4.4 The Regionalization of Manufacturing Industry Chain and Service Industry Chain

In the future, driven by technology, market, government and cost, the regional attribute of efficiency-driven value chain of manufacturing industry chain, especially automobile, electronic products, textile and garment, will be further enhanced, while the global attribute of service industry will be further strengthened. For labor-intensive industrial chains such as textile, clothing, toys, shoes and furniture manufacturing, with the rise of labor costs in developing countries and the substitution of automation and artificial intelligence technology for labor, the arbitrage space based on labor costs will be further reduced, or it will be transformed into a capital-intensive industrial chain and shrink to the final consumer market or the region where the home country is located. Human capital, technology and government are the main factors driving multinational corporations to expand their supply chains in the capital-and technology-intensive manufacturing industries such as chemical industry, automobile, computer, electronics and mechanical equipment. Automation technology expands the location space of labor-intensive processing and assembly links. With the characteristics of long production chain, large number of suppliers, close production process and sensitivity to external shocks, automation technology prompts multinational companies to achieve production requirements on time and reduce the social cost and coordination. For metal products, rubber and plastic, paper and printing, food and beverage and other industries, due to the weight, volume and perishable characteristics of products, regional characteristics are relatively significant, about two-thirds of the products as intermediate products into the capital and technology intensive industry chain, especially the regional industry chain. For the industry chain of resource-intensive products such as agriculture, mining and base metals, natural resources and transportation convenience are still the main determinants of site selection. Meanwhile, unlike some manufacturing industries driven by the digital economy, the service industry will achieve further globalization with more detailed industrial division and more widespread service outsourcing.

5.5 Automotive Industry Development with Industrial Chain Reconstruction

The global spread of COVID-19 has had a huge impact on the supply of raw materials, the supply of factors of production, the production of intermediate goods and the final consumer market. As a typical capital- and technology-intensive industry, the automobile industry is more severely impacted by the pandemic due to its high technology content, fine labor division and long industrial chain. The refined vertical division of labor has led to the vulnerability of the global industrial chain system and the failure of any component supply will have a serious impact on the production of upstream and downstream countries. In the early stage of the pandemic, China's supply of spare parts was cut off, triggering a "butterfly effect" around the world. In February 2020, affected by the interruption of auto parts supply from China, many auto companies in South Korea and Japan stopped their production one after another. Just four days from March 13 to 16, factories of seven brands, including Ford, Volkswagen, Renault and Ferrari, announced temporary closures. As the global "auto processing plant", China still relied on imports from developed countries for key components, and the suspension of production by overseas auto companies has also had a serious impact on Chinese auto companies.

The impact of the pandemic on the global automotive industry chain can be seen from changes in automobile production in major auto producing countries such as the United States, Germany, Japan, South Korea and China. As shown in Fig. 5.1, the automobile output of major automobile producing countries showed negative growth to varying degrees in 2020. Among the 12 major auto producing countries, China's auto industry as a whole has been least affected by the pandemic, followed by South Korea, while France's auto production has been hardest hit by the pandemic, down nearly 40% from 2019. East Asian countries have been relatively less affected by the virus, while European countries have been hardest hit by the virus in the Northern United States.

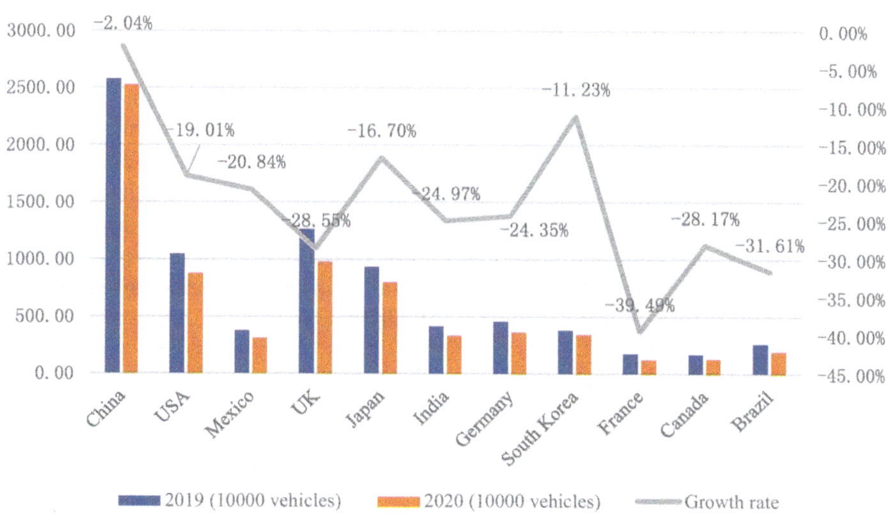

Fig. 5.1 Changes in production and growth rates in major automotive countries in 2019 and 2020. *Source* OICA, www.oica.net

5.5.1 *The Restructuring of Global Automotive Industry Chain*

Many factors determine the reconstruction of the global industrial chain. Multinational corporations' pursuit of modularization in production process, benefits of specialization and economies of scale lead to longer value chain, while the increase of trade cost and innovation intensity will shorten the length of value chain. A lot of arbitrage opportunities and the high degree of product customization make the geographical distribution of added value more dispersed. The increase of supply and demand concentration and trade and transportation costs will lead to the geographical distribution of added value more concentrated. According to UNCTAD's World Investment Report 2020, the new industrial revolution, economic nationalism and sustainable development have become three trend factors driving major changes in the global industrial chain in the past decade. These trends sometimes reinforce each other, and sometimes push the evolution of the global industrial chain in opposite directions, making the evolution patterns of various industries and regions different. For example, the development of robotics, 3D printing, cloud computing as well as other technologies can flatten, squeeze or bend the "smile curve" of international production, exerting significant influence on the length, geographic distribution and governance patterns of global value chains. Economic nationalism reduces the scale of international direct investment, goods trade and global value chain trade. Specifically, state interventionism in industrial policy supports the concentration and agglomeration of knowledge and technology in GVCS, integrates modular value chains and offsets labor cost arbitrage opportunities. Trade and investment protectionism measures, such as raising trade barriers, strengthening foreign investment review and restricting outbound investment, hinder the labor division and geographical diffusion of the value chain. The shift from multilateral to regional and bilateral policy frameworks has reduced the cost of intraregional trade and enhanced indigenization and the development of regional value chains. Sustainable development issues affect the behavior and governance choices of MNCS. Regulatory policies, consumer preferences and reputational risks are changing the behavior of MNCS, driving the shortening and stricter control of supply chains.

5.5.2 *Environmental Factors*

As for the automobile industry, the macro-environmental factors that affect the reconstruction of global automobile value chain can be divided into three categories, economic factors, technical factors and political factors. Economic factors involve labor cost, trade cost, transportation cost, transaction cost and other cost factors, as well as modular production, scale economy, supply and demand concentration, product customization degree and green economy requirements. Technical factors involve automation, digitalization and other technologies. Political factors, including various incentive policies and restrictive policies, also have an important impact on the global production layout of multinational automobile companies.

5.5.2.1 Economic Factors

Economic factors mainly influence the length of automotive global value chain and geographical distribution of added value through the following aspects. First, the outsourcing of production services by MNC is hindered by the reduction of arbitrage opportunities. The vertical division of the value chain is deepening. Refined production processes lead to rising trade costs, transportation costs and transaction costs, making the global value chain develop in the direction of shorter length and more concentrated geographical distribution of added value. Second, multinational companies' pursuit of short-range supply and scale economies in integrated production process has led to concentrated automobile production layout, shorter length of global value chain and more concentrated geographical distribution of added value.

In addition to the economic factors above, green economy, which determines the trend of global value chain reconstruction, cannot be ignored. The rise of new energy vehicles and the global green economy is closely related to the theme of sustainable development. In order to achieve sustainable development, all countries government policies and measures in specific industry increased significantly. The trend of sustainable development and the advocacy of green economy led to the transformation from traditional fuel vehicles to new energy vehicles, which requires all countries to promote the development and application of energy efficiency technology. Undoubtedly it brings great changes to the automobile industry chain all over the world.

5.5.2.2 Technical Factors

The influence of technological factors on the reconstruction of global automotive value chain is mainly reflected in the application of automation technology and digital technology. Automation relies on the use of advanced robots, whose penetration in the automotive industry is very high and expected to grow in the future. These advanced industrial robots have the computing power to replace low-skilled labor workers, making it possible for the auto manufacturing industry to move back. Digital technology can improve the integration of production processes, reduce transaction costs, improve the collaborative efficiency of automotive industry chain and promote the short-chain of global value chain. The use of digital technology also reconfigures the added value of each production link. Digital technologies such as the Internet and big data highlight the importance of intangible assets in the value chain, further reducing the value-added content of intermediate processing and manufacturing links and making the distribution of value-added more concentrated at both ends of the value chain.

5.5.2.3 Political Factors

Political factors also play an important role in the reconstruction of GVCS. The COVID-19 pandemic has further strengthened the dominance of economic sovereignty and fear of the free trade system, greatly shaking the political foundation of western countries supporting economic globalization. During the pandemic period, main western countries are competing to introduce restrictive industrial policies, strengthen export control and domestic strategic industrial protection, tighten foreign capital security review, encourage industrial relocation, strive to improve the national or regional supply capacity of intermediate products and final products, and realize self-sufficiency and self-control of the economy. There has been a significant inward transformation trend in industrial policies. In early March, French Finance Minister Jean-Pierre Le Maire said it was necessary to reduce dependence on certain countries, especially China, for supplies of products such as automobiles, aerospace and pharmaceuticals, and strengthen sovereignty in strategic value chains. The United States announced the implementation of the *Emergency Act and the Defense Production Program*, proposing that all material security must be subject to the Goal of America first, and the key industrial chain must be moved back to the United States. In early April, the Japanese government planned to spend 220 billion yen to help Japanese companies move production back to Japan and 23.5 billion yen to help Japanese companies move production to other countries. The German government amended its *Foreign Trade and Payments law* to prevent foreign investors from taking advantage of the opportunity to acquire companies hit hard by the pandemic, such as auto companies.

These policies have a significant impact on the global automotive production network layout by changing the mobility of factors of production and final products. Incentive policies promote the mobility of factors and products, while restrictive policies hinder the mobility of factors and products. When factors and products can flow freely in the market, automobile enterprises will carry out production layout according to the resource endowment of each country or region and comprehensively consider economic factors such as cost, market and technical factors. On the contrary, the influence of economic and technological factors on production layout is weakened, while the decisive role of political factors on production activities is strengthened, which undoubtedly has a significant impact on the location choice and production layout of automobile multinational corporations.

The policy shift of western countries is accelerating the trend of global automotive value chain reconstruction. Through policy adjustment, the developed countries in the automotive industry led by the United States and Germany will adjust the global layout of the automotive value chain to some extent, and tend to build a more independent, more complete and safer industrial chain, which is bound to have a profound impact on the global automotive value chain. The withdrawal of western countries from the automotive supply chain will seriously affect the production cooperation of upstream and downstream countries and the policy of encouraging enterprises to move back is likely to be the catalyst for the further contraction and localization of the supply chain of domestic enterprises.

5.5.3 Global Automotive Industry Chain in the Post-pandemic Era

In the post-pandemic era, the safety of the automotive supply chain will be more important than efficiency, and self-sufficiency in key core links will become a new trend. Automotive multinationals will be forced to rethink their global value chains to better balance efficiency and resilience. Combined with economic, technological and political factors, the global automotive industry chain will be in a critical period of transformation in the next decade. Although there is considerable uncertainty about the extent and scope of the transformation, some important trends have emerged.

5.5.3.1 Short Chain

As labor costs, trade costs, transportation costs and transaction costs continue to increase, the degree of product customization and sustainable development requirements, the global automotive value chain will develop in a shorter direction. Driven by sustainability trends, production and investment in the auto industry is likely to shift to electric vehicles over the next decade, with some countries already setting a goal of phasing out internal-combustion engines by 2030. With fewer components and a shorter value chain, the added value of the auto industry will be more concentrated.

Traditionally, GVCS in the automotive industry have been highly intensive, with complex transnational production networks around the world and suppliers in many regions. The shift to electric vehicles will change existing production patterns and lead to a restructuring of GVCS. The powertrain of a conventional internal-combustion engine has more than 2,000 moving parts in a conventional fuel-powered car, while the powertrain of an electric car has only 20. The added value distribution will be concentrated in a few parts, such as the battery, central control and chip. The number of suppliers would also be greatly reduced, shortening the global value chain for cars. Tesla, for example, has only about 300 suppliers in a handful of countries, while most traditional automakers have thousands of suppliers worldwide, with BMW 4,500 and Nissan 5,000.

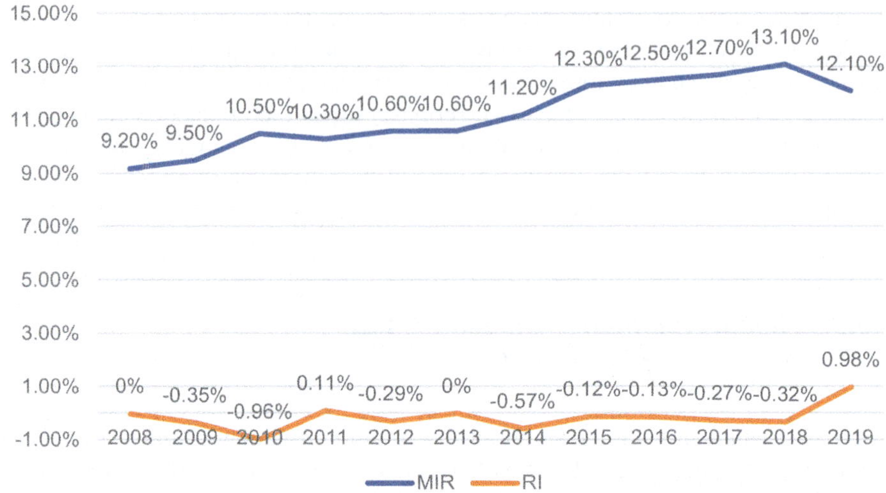

Fig. 5.2 US Manufacturing Import Ratio (MIR) and US Manufacturing Reflux Index (RI). *Note* US manufacturing import Ratio (MIR) = total imports of manufactured goods ÷ total output value of domestic manufacturing × 100%. Among them, the total amount of manufactured goods imports is the total amount of imports from 14 Asian countries (regions) such as China, Taiwan, Malaysia, India, Vietnam, Thailand, Indonesia, Singapore, the Philippines, Bangladesh, Pakistan, Hong Kong, Sri Lanka and Cambodia. RI = Ratio of US manufacturing imports in the previous year–Ratio of US manufacturing imports in the current year. If RI is positive, it indicates that manufacturing is coming back. *Source* US International Trade Commission, Bureau of Economic Analysis, US Department of Commerce

5.5.3.2 Localization

In the post-pandemic era, driven by the return trend of manufacturing industry in developed economies and the policy environment to enhance industrial capacity, the localized development of automotive industry chain will be very common. As shown in Fig. 5.2, before 2018, the US manufacturing import ratio showed a trend of increasing in general. In 2019, it was 12.1%, declining for the first time since 2011. Before 2018, the manufacturing backflow index in the United States was basically negative. In 2019, there was a surprising reversal. Compared with 2018, the backflow index increased by 98 basis points, indicating that the manufacturing output in the United States increased substantially and the trend of manufacturing backflow in the United States was accelerating.

5.5.3.3 Regionalization

Before the pandemic, the trend of regionalization of the automotive industry chain was intensifying in East Asia and North America, such as the return of auto parts production to the US-Canada-Mexico Free Trade Zone. The outbreak of COVID-19 has accelerated this trend. With the background of the return trend of manufacturing industry in developed economies and the deepening of regional economic cooperation, the automobile industry has begun to shift from global value chain to regional value chain. The pandemic has prompted MNCS to shift their investment to neighboring countries with trust, reducing their dependence on a single country and thus diversifying risks to ensure the safety of supply chains. The regionalization nature of product trade is constantly strengthening, especially in Asia and Europe. Automobile enterprises are more and more inclined to conduct production activities in regions adjacent to consumer markets, and the investment motivation of enterprises is gradually shifting from cost-income seeking to market seeking. Technically, digital development, including digital technology and digital infrastructure, is conducive to the centralized coordination of regional value chains and provides the possibility to replicate the whole industrial chain within the region.

5.5.3.4 The Distribution of Added Value

In the context of the vigorous development of the "new four modernizations" of automobiles, enterprises pay more attention to front-end R&D and design and back-end software development. The proportion of added value in intelligent technology, battery technology and service links increases, while that in traditional production and manufacturing links decreases. With the new energy emerging, electric drive system will become the main growth point of vehicle value. Battery, electric drive and electric control are the three core technologies of new energy vehicles, and battery and power system account for more than half of the total cost of pure electric vehicles, so the added value of R&D and design has significantly increased. In addition, with the development of digitization and informatization downstream of the industrial chain, the added value of post-market services, such as car sharing and autonomous driving, will account for an increasing proportion. The content of added value at both ends of automobile value chain is further increased, the content of added value at the intermediate manufacturing link is further reduced and the smile curve is steeper, which will become an important form of automobile value chain distribution.

5.5.3.5 Internalizing the Governance Pattern

The internalization trend of governance mode of global automotive industry chain will be further accelerated, with the decrease of offshore outsourcing and subcontracting, the increase of onshore or near-shore insourcing and the deepening of vertical integration within industrial clusters. In the post-pandemic era, the internalized governance mode of the automotive industry chain may play a dominant role, especially the tendency of internalization of knowledge-intensive activities. There are two main purposes for the internalization of automobile multinationals. First, the internal organization can replace the external market to reduce transaction costs. Multinational companies seek to gain more revenue by internalizing and integrating intermediates from research and development to production and assembly. Second, the market incompleteness can be alleviated and buyer's uncertainty can be eliminated through intra-enterprise transaction to reduce transaction risk. Due to the particularity of knowledge-based products and the market failure caused by moral hazard, information asymmetry, incomplete market and other factors, multinational enterprises will internalize the transaction activities of knowledge-based products in order to reduce risks. In fact, not only the automobile industry, but also more asset-light, relationship-heavy and knowledge-heavy enterprises will occupy a leading position in the global industrial chain of technology-intensive industries, and the R&D activities of core technologies are becoming more and more internalized.

5.5.4 Global Auto Industry Chain Reconstruction

The restructuring of the global automotive industry chain had a great impact on the economic development of developing countries including China and their participation in the international labor division. The development trend of short chain, localization and regionalization means that the international production layout of multinational corporations will be close to or return to developed economies, reduce direct investment and value acquisition in developing countries, block their access to advanced technology, and make the export-oriented development strategy and industrial transformation of developing countries based on vertical specialization severe challenges. The distribution of added value is shifting to both ends of the value chain at an accelerating pace, and the governance model of global industrial chain tends to be internalized, which means that China's participation only in the process of manufacturing is not conducive to the promotion of its position in the global value chain.

5.5.4.1 Development of Chinese Automobile Industry

At present, China's automobile industry has been deeply integrated into the global automobile value chain, but it shows weakness in key technologies and core components. On the whole, the global value Chain (GVC) status index of China's automobile industry was not high from 2000 to 2014. It showed a downward trend before 2005 and then a slow upward trend, and exceeded 0.1 in 2014 (see Fig. 5.3). This is mainly because the high value-added links in the automotive value chain, such as basic R&D system design, key components R&D and design, brand management, distribution and after-sales service, were basically provided by the parent companies and subsidiaries of multinational companies, while Chinese enterprises were mostly engaged in low and medium value-added links (production of parts, assembly, logistics, etc.).

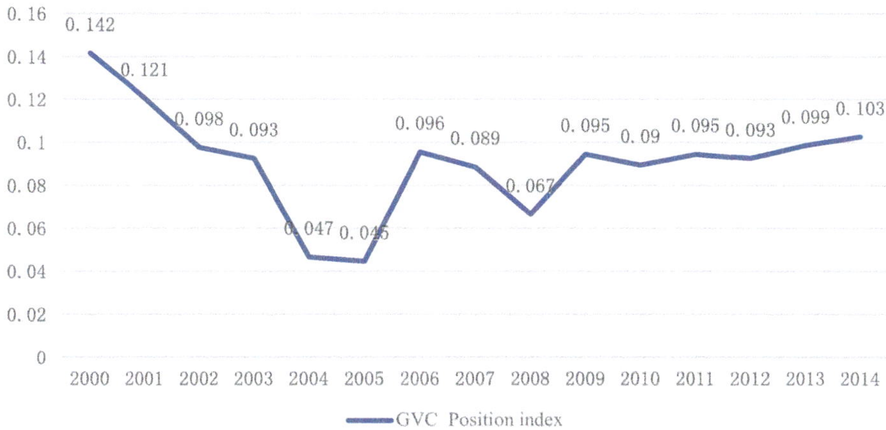

Fig. 5.3 GVC status index of China's automobile industry from 2000 to 2014. Note: In this paper, C29 in the Input-output table of WIOD database is used to represent the automobile industry, and the GVC status index is calculated according to the method proposed by Koopman et al. as $GVC_{Position_{ik}} = Ln(1 + IV_{ik}/E_{ik}) - Ln(1 + FV_{ik}/E_{ik})$ where E_{ik} means the total exports of k industry in country i to other countries in the world calculated by added value and IV_{ik} represents indirect value-added exports of k industry in country i. FV_{ik} is foreign added value in exports of k industry of country i. Since the latest data of WIOD was updated to 2014 (released in 2016), this paper calculates the GVC status index of China's auto industry from 2000 to 2014. *Source* Calculated according to the input-output table in WIOD database

China's auto industry has a close relationship and high dependence on global auto value chain. Germany, the United States and Japan are the world's major auto exporters, while China has the world's largest auto production base. As shown in Fig. 5.4, China needs to import a large number of automobiles and parts from developed countries and regions in the automobile industry such as Western Europe, the United States and East Asia every year. In 2020, the import amount from Germany, Japan and the United States alone accounted for 67.69% of China's total import of automobiles and parts.

5.5.4.2 Impact of Global Auto Industry Chain Reconstruction on Chinese Auto Industry

First, the space for China's auto industry to participate in the division of global auto value chain has been further compressed. The trend of short chain, localization, regionalization and internalization of governance mode of global automobile industry chain reduces the opportunities for Chinese automobile enterprises to participate in the division of global value chain. A large number of automobile multinationals from developed countries may transfer their production in China, and Chinese enterprises will have less chance to participate in global automobile value chain. During the

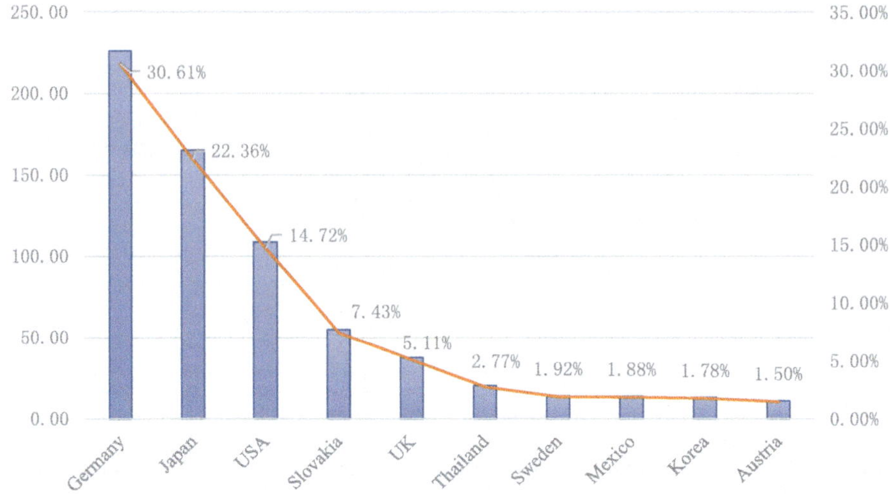

Fig. 5.4 Import dependence of China's auto industry on various economies in 2020. *Note* Import dependence of automobiles and parts = a automobiles and parts imported by China from one country ÷ total import of automobiles and components × 100%. *Source* UN Comtrade Database

pandemic, Mazda Motor Company significantly increased its components production in Mexico and Japan. Komatsu has shifted production of metal parts and electronic wiring harnesses used in car bodies from China to Japan and Vietnam and the shift of automotive supply chains from China has adversely affected Chinese auto companies' participation in the international labor division.

Second, the restructuring of the global automotive industry chain has blocked China's access to important sources of core technologies. China is a big country in automobile manufacturing, but not a strong one. In terms of core technologies and key components, China relies heavily on imports from developed countries. In the field of new energy vehicles, although China's production and sales of new energy vehicles rank among the top in the world, most of them mainly rely on domestic sales, and domestic automobile brands are not competitive with world famous automobile brands. The reconstruction of the global automobile industry chain has greatly reduced the dependence of developed countries on China, which is difficult for China's automobile industry to deeply integrate into the global automobile value chain, making it harder for China's automobile enterprises to learn foreign advanced technology. In addition, China's automobile enterprises lack the research and development of key new technologies, so it will undoubtedly be more difficult to climb to the high end of the value chain.

The absence of international well-known automobile brands leads to the loss of the right to make industrial rules in China. The lack of discourse power is an important reason why China's automobile industry cannot occupy a dominant position in the global value chain labor division system. In 2019, the world's top 10 most valuable automobile brands all belong to Germany, the United States and Japan, and most of them are traditional car companies such as Toyota, Mercedes, BMW, Honda and Ford. Tesla is the only electric car company on the list. With the transformation from traditional fuel vehicles to new energy vehicles, a new round of international rules for automobiles is being constructed. Electric vehicles represented by Tesla are taking advantage of their own advantages and seizing the dominant position in the industry. The international popularity of China's own brands is not high, facing great pressure in the new round of international competition. Of course, the restructuring of the global automotive industry chain also provides new opportunities for China's automotive industry to reach the world-class level. European and American automobile leading enterprises rely on the monopoly of fuel car core technology, in the world's top position. Compared with engine and transmission technology of fuel vehicles, the complexity of core technologies such as battery, armature and ECU of electric vehicles is greatly reduced, providing opportunities for new enterprises to enter the automotive industry. After more than a decade of development, China's new energy vehicle industry has begun to take shape. In the first half of 2019, 563,000 new energy passenger vehicles were sold in China, accounting for 56.9% of the global market share, far higher than the EU's 20%. With the application of digital and intelligent technology, such as the Internet giant Tencent, BaiDu with its technical advantages, strengthen the research of automatic driving technology, and traditional Chinese automakers intensify joint innovation, is expected to be in power system and intelligent driving a breakthrough, in order to enhance the international competitiveness of China's auto industry.

5.5.5 Development Strategy and Policy of Chinese Automobile Industry

In the post-pandemic era, the restructuring of the global automotive industry chain will have a significant impact on the development of China's automotive industry. At the same time, it also brings some development opportunities for China's automobile industry to realize the high-end leap. China urgently needs to coordinate development and safety, establish a more flexible, sustainable and safe automobile industrial structure and speed up the construction of domestic and regional industrial chains. Internally, we will strengthen industrial safety and technological independence, accelerate the promotion of "chain complement", "chain expansion" and "strong chain", and strive to build a domestic demand-oriented domestic automobile industry system. Externally, we should embed the strategic path of the global automotive industry chain and participate in international cooperation and competition in a more resilient and sustainable way.

First, we need to accelerate industrial chain upgrading. Through scientific and technological innovation, research and development will be carried out to target the weak links and high-end links of the industrial chain, and core technologies will be accelerated to realize the high-end of the industrial chain and overcome the "low-end lock-in" as soon as possible. We will vigorously promote the digital, intelligent and green development of the automobile industry, and accelerate the intelligent upgrading of automobile production methods by utilizing artificial intelligence, cloud computing and industrial Internet technologies. Focus on new energy vehicles, smart connected vehicles and other areas, seize the dominance of technology and product standards, and strive to occupy a leading position in the new round of global automotive industry changes.

Second, to build an automobile industrial cluster, China needs to take the core enterprises of the automobile industry chain as the leader, optimize the industrial supporting environment and conditions, accelerate the coordinated development of upstream and downstream of the industrial chain, production, supply and marketing, large and medium-sized enterprises, strive to build a number of world-class industrial clusters with high upstream and downstream coordination, close technical connection and intensive and efficient process, and enhance the new advantages of international cooperation and competition of the automobile industry.

Third, to adjust the mode of utilizing foreign capital, China should unswervingly promote the strategy of opening up, effectively improve the quality of utilizing foreign capital in the automobile industry, and shift the focus of foreign capital policy to "attracting capital to complement the chain", "attracting capital to expand the chain" and "attracting capital to strengthen the chain". China will encourage Chinese and foreign enterprises to jointly conduct research and development on basic research, applied research and generic key technologies, and actively foster new technologies, new products, new business forms and new models of the automobile industry.

Fourth, to implement a new round of "going global" strategy, China should actively encourage and guide enterprises to enter the markets of developed countries, gradually embed the automotive global innovation network dominated by European and American enterprises through mergers and acquisitions of high-tech projects with core technologies, joint development and joint establishment of overseas R&D centers, so as to expand the international competitiveness of Chinese enterprises' Industrial Brands. China should continue to promote "follow-up" foreign investment, closely track the strategic trend of multinational corporations, and actively strengthen their supporting and outsourcing relations and the upstream and downstream relations of the global automobile industry chain with capital as the link.

Another thing to do is strengthening international cooperation on industrial chain. China should promote international cooperation among ecological players in the industrial chain, especially with Japan, the ROK, ASEAN and other neighboring countries, step up joint research and development and design, core technology innovation, marketing development, parts upgrading, high-level brand cultivation, and jointly build and share emerging consumer markets. What should be done includes giving full play to the role of the Belt and Road industrial cooperation platform, expanding the export and investment activities of auto products to countries along the Belt and Road, and upgrading the level of industrial cooperation with countries along the Belt and Road.

References

Bullock, Charles J. 1902. The Variation of Productive Forces. *The Quarterly Journal of Economics* 16 (4): 473.

Li, Jing and Houkai Wei. 2007. The Clusterization Strategy of Chinese Industrial Zones Based on Industry Chain. *Economic Survey* (2): 68–71.

Liu, Gang. 2005. Structure of Knowledge Transitions and Innovations Based on an Industrial Chain. *Journal of Business Economics* (11): 13–17.

Liu, Mingyu, and Jin Weng. 2007. Division of Industry Chain and Knowledge Integration. *Science of Science and Management of S&T* 28 (7): 92–96.

Lu, Minghua and Guoping Li and Xiaobing Yang. 2004. The Development of China's Electronic & Information Technology Industry in View of Industry Chain. *Forum on Science and Technology in China* (4): 18–22, 96.

Marshall, L.C. 1913. Incentive and Output: A Statement of the Place of the Personnel Manager in Modern Industry. *Journal of Political Economy* 28 (9): 713.

Stigler, George J. 1942. The Extent and Bases of Monopoly. *American Economic Review* 32: 1.

Wang, Yunxia and Guoping Li. 2006. Review on the Current Situation of Industrial Chain. *Journal of Industrial Technological Economics* (10): 59–63.

Xu, Jinan. 2007. A Study of the Strategy of the Industrial Proprietary Brand Based on Industrial Chain. *Contemporary Finance & Economics* (2): 71–76.

Yu, Yihong, and Xizhan Guan. 2006. *Vertical Control of Industrial Chain and Economic Regulation.* Shanghai: Fudan University Press.

Zhao, Hongyan. 2008. The Integration of the Industry Chain and the Development of China's Enterprises. *Contemporary Finance & Economics* (9): 78–83.

Chapter 6
China's High-Tech Industry and Global Industrial

This chapter is mainly based on how countries participate in the restructuring of GVCS in the context of the global spread of COVID-19. In the first part of this chapter, three ways to participate in the reconstruction of global value chain are proposed, which are embedding in global value chain, constructing national value chain and leading regional value chain. The second part judges which path Chinese industry should choose to participate in global value chain reconstruction according to VRCA index. The third part elaborates the influence of "One Belt and One Road" initiative on China's industrial upgrading, how to support the construction of regional value chain of China's industry, its effect on neighboring countries, and puts forward the direction of future development of China's industry. The last part, based on the "Belt and Road" initiative, analyzes the feasibility and development potential of China's photovoltaic industry to build the regional value chain along the "Belt and Road", and proposes an effective path for the development of photovoltaic industry.

6.1 Paths of Participation in GVCs Reconstruction

China's high-tech industry has a non-negligible disadvantage in Global Value Chains (GVCs) division of labor and there is a risk of "low-end locking" when participating in GVCs division of labor. Many intellectual property rights related to design in China's high-tech industry are still controlled, making most of the core values obtained by other enterprises outside China. The sudden global spread of COVID-19 has disrupted the cyclical process of the world economy and made the disruption of the global supply chain of China's high-tech industry. With the development of new-generation information technologies, such as mobile Internet, cloud computing and big data, the manufacturing and organization modes of some industries, especially high-tech industries, have undergone major changes, prompting the decomposition, integration and innovation of value chains on a global scale. The "smile curve" of international industrial division of labor has been seriously deformed and the added

Y. Jiang and J. Gu, *Technology and Industrial Transformation of China*,
https://doi.org/10.1007/978-981-19-7458-8_6

value of all links has changed accordingly. GVCs are undergoing a new round of reconstruction and the value chain is gradually extending to a higher level of new economy and chain. The high-tech industry represented by AI, 5G, smart Logistics and online payment has begun to take shape and its international influence is gradually expanding. China needs to grasp the new opportunities of technological change, advance into high-end manufacturing and actively participate in the restructuring of global value chain.

The global value chain concerns the selection and connection of upstream and downstream links of products. With the fragmentation of labor division and the diversity of restructuring methods, the selection and connection of upstream and downstream links of products also have a variety of combination paths, forming a variety of value chains. The network structure formed by the combination of multiple value chains can adapt to changes and respond to customer preferences quickly, which can bring enterprises into an intertwined, more large-scale and complex enterprise group for research. In recent years, the concept of "Value Web" has been extended to explore the virtual complex division of labor and collaboration system without asset bond. With the development of globalization and industry, there will be unlimited possibilities for the value-added ability of the industry in the future and the boundary of links will no longer be clear. The global value chain is not a chain, but consists of different forms of GVC. For example, Diakantoni et al. (2017) and Balwin et al. discussed the organizational form of global value chain, based on the geographical location of supply chain. They divided the organizational form of global value chain into spider-shaped value chain with hub and spoke-structure distribution, snake-like value chain and mixed value chain.

The GVCS shown in Fig. 6.1 have a network structure. The value chain in the global value chain network has the following three characteristics, different forms, varying in length and interlaced with each other, these chain forms the value network that constitutes the global value chain system. Chain 1 is the traditional form of a certain type of value chain and F is the enterprise on the value chain. There are several manifestations of value chain reconstruction. Along the direction of the original value chain, it moves to the high end, as shown in the figure F moves to F1, enhancing its value creation ability. Based on the expansion of the integration of industry and technology, the traditional "smiling" morphogenesis deformation value chain, the chain 2 is the new mode of value chain after changing. Although the enterprise comparative advantage and competitive advantage of F has not changed, because of its profitability segment difference, its position in the new chain F2 is better than the original chain. In the original global division of labor, enterprise F obtains the technology spillover effect from the enterprises of the investing country. However, in order to reduce the dependence on foreign technology and gain greater innovation capability, enterprise F tends to establish the national value chain F4 to enhance the independent innovation capability and control the resources of the country. If the country of enterprise F has a relative competitive advantage in the international competition of the industry, its position in the original value chain division is not high and the profit is low. Despite the local value chain has been built, "independent innovation nationalism" technology brings tension in international relations. It requires a

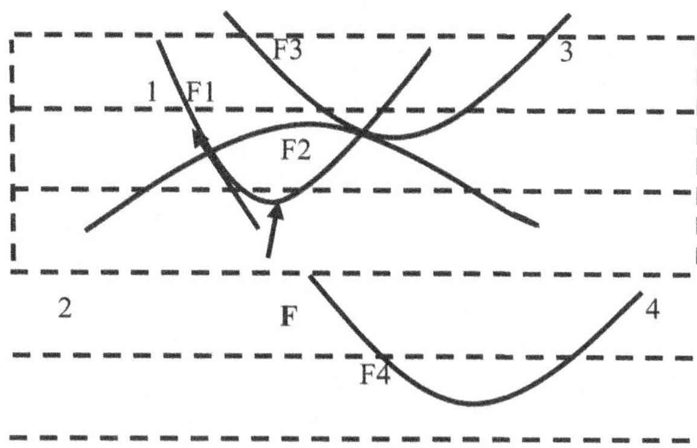

Fig. 6.1 A country's participation in the global value chain restructuring path framework

combination of different countries in this industry competitive and complementary relationship, establish a new contact area. In the process of regional chain transformation, its competitive advantages are gradually developing and changing. It is possible to re-participate in the new global value chain division system by virtue of its new comparative and competitive advantages in the future and move to F3 by participating in the new division of labor in chain 3 with a new role.

To sum up, Fig. 6.1 provides an analytical framework for a country's enterprises to participate in GVC reconstruction. The first path is embedded GVCs. Enterprise F moves to F1 in chain 1 to enhance its value creation ability or improve its position in the value chain through industrial integration, technological expansion and chain form changing. The second path is to build NVCs. F's upward path from chain 1 to F1 is blocked and its independent innovation ability is enhanced to construct NVCs on chain 4 at the position of F4. The third path is to create the dominant RVCs. With the integration of industries and expansion of technologies, F tends to choose chain 3 by taking advantage of new competitive advantages and create dominant RVCs on chain 3 at the position of F3. The three paths are different in chain form, industrial development characteristics, stages and path selection requirements.

6.1.1 Global Value Chains

Developing countries crossing the middle-income trap should develop labor division from trade specialization to technology specialization. The first step for enterprises is to learn on the basis of licensing, transfer or FDI and to establish preliminary technology absorption capacity. Domestic enterprises can realize the development from OEM to ODM through FDI. The path embedded in global value chain discusses moving to the link creating higher value, producing products with higher value and improving production methods through production process innovation or new technology introduction from the original value chain. The purpose is to pursue high productivity, which is embodied in extending to the upstream of value chain and the downstream of the value chain. A country's enterprises learn and imitate in the technology demonstration of multinational corporations, concentrate on the business activities of a certain core link, outsource other non-core links, promote the industry to move upstream along the value chain, and change the profitability and the country's position in the international labor division. As for extending to the downstream of the value chain, we can see what was pointed in the *UK Industry 2050 Strategy*, modern industry, especially high-tech industry, is not only including manufacturing and then selling, but also "service and remanufacturing". Enterprises' labor cost advantage is no longer unable to gain a competitive advantage in the middle segments. However, they can rely on the service cost advantage to carry out a new round of business behavior, enter the downstream sales and after-sales links of the industrial chain with the new competitive advantage, and even have the trend of extending to the downstream links so as to occupy the initiative of more links in the value chain. No matter whether they move to the upstream or downstream of the value chain, countries and industries that choose to actively embed themselves in the global value chain must have strong competitiveness and profitability in the global value chain.

6.1.2 Building National Value Chain

GVCs labor division is helpful for developing countries to import high-tech intermediate products from developed countries and obtain R&D achievements and technology spillover effects. However, due to path dependence and status solidification, it is difficult for embedded value chain to change the labor division status of value chain. The deep integration of GVCs has brought trade development at high-speed, but when analyzing the essence of overall trade profits, countries at the high end of the value chain have controlled the whole value chain and the global pricing power of products. Its utilization technology monopoly advantage hindered the value chain of low-end countries along the value chain to high value-added links up, even the implementation of "technology lock". In the global value chain division system, the interests of enterprises at the low end of the value chain are exploited and their development is limited. There were great obstacles in the natural upgrading process. The markets of developing countries or emerging economies were controlled by foreign companies and most core technologies depended on imports. In the process of function upgrading and chain upgrading in the original value chain, part of China's high-tech industry was still locked in low value-added production links or directly eliminated from the global division of labor system. According to this situation, national policy encouraged to temporarily slow down the integration of the global value chain, giving priority to improving the added value of products at home and replacing imports, and gradually developed and cultivated their own high-level elements through accumulating learning and imitation experience. The main path is to build the domestic value chain, through the value chain in China upgrading to drive the improvement of industry status in the international labor division, completing value chain reconstruction.

Backward countries should make a dynamic change from the traditional industrial policy at the initial stage of development to the technology policy at the later stage of development. Domestic enterprises should set up their own internal research centers to focus on their internal capabilities, and establish overseas R&D cooperation with foreign companies or public institutions to initiate international mergers and acquisitions. In 2006, emerging economies represented by China began to reduce their dependence on foreign technology, increase their control over intellectual property rights, and carry out major policy adjustments of "independent innovation". After developing countries or economies achieved process upgrading and product upgrading through low-end embedding into the global value chain, local enterprises mastered the core links of the value chain such as R&D, brand and marketing, key links of the value chain, key resources and obtained market demand. The construction of national value chain can win the favor of enterprises in emerging industries more than integrating into the global value chain. After low-end embedding into global value chain, the cultivation of national value chain can transform the capture network into a balanced network, re-integrate the industrial association and circulation system that a country's enterprises rely on for survival and development, and reshape the national governance structure and the association structure between regional industries. Then accumulating enough advanced factors of production, it can extend from the national value chain to the high end of the global value chain to seek the dominance of the global value chain.

6.1.3 Regional Value Chain

The leading regional value chain is a regional cross-enterprise network organization that selects a specific region to form agglomeration effect, combines with emerging countries or regions with strong complementary industries around them and connects the whole industrial chain to realize the value of goods or services, so as to realize industrial upgrading and create more value. For enterprises in developing countries, stepping into industries with short cycle is the main scheme to complete the status leap, so the choice of industries is particularly important. With the penetration of the Internet and the integration of information technology, many high-tech industries have begun to undergo transformation. These transformed emerging industries are more suitable for enterprises to make leaps and bounds. Compared with the traditional industrial value chain form, the new generation of information industry, intelligent manufacturing, represented by large-scale intelligent customization and "Internet plus", their value chain form has changed. Industrial chains tend to be flat and evolution trend of central uplift. The industrial convergence phenomenon makes the enterprise more technology and capital accumulation.

Although the development of national value chain is the policy requirement of "independent innovation" by developing countries or emerging countries, the connection between R&D process and industrialization process is weak and the process of scientific and technological achievements is relatively long. In the long time, "nationalism" technology may cause tension in international relations and slow the speed of technological progress. Therefore, while it drives the construction of national value chain, independent innovation must also fully consider the effective connection with the international value chain. In fact, with the deepening of global value chain division, the trend of value chain regionalization has become more and more obvious, and has formed a regional value chain dominated by the US and Europe in the service sector, Japan and China in the manufacturing sector, and other countries in the field of natural resources.

6.2 Chinese High-Tech Industry in Reconstruction

6.2.1 Selection Basis

Which path should China choose for different types of high-tech industries to participate in GVC reconstruction? In the process of path transformation, will high-tech industry still embed in global value chain, build national value chain, or dominate regional value chain? Do different industries have the same choice? Whether it is to embed in global value chain, construct national value chain or lead regional value chain, there are certain requirements for the behavior of enterprises, industrial characteristics and national strategy, the path choice depends on the characteristics of the industry and the international competitiveness of the country and the industry.

According to the theoretical analysis above, there are three ways for a country's industries to participate in the reconstruction of global value chain. In terms of the differences in competitiveness, China's high-tech industries can choose three ways to participate in the reconstruction of global value chain as shown in Table 6.1.

How to measure the international competitiveness of China's high-tech industry? With the increasing degree of vertical separation and fragmentation of value chain, the traditional RCA index ignores both domestic division of production and international production sharing in the context of global value chain division, and fails to reflect the real competitive advantage of export industry. Therefore, this paper refers to the modification of RCA index by Wei Long and Wang Lei, Li Huiru and Chen Zhaowei, and establishes the evaluation index system of international competitiveness of China's high-tech industry. Through the value-added trade analysis framework, VRCA (value-add Revealed Comparative Advantage) index is used to measure

Table 6.1 Paths for China's high-tech industry to participate in the reconstruction of global value chains

Paths	Chains	Competitive power
Embed in GVC	GVC	Extremely strong
Build NVC	NVC	Strong
Dominate RVC	RVC	Relatively weak

the dominant comparative advantage index of different industries, and path selection proposed for different high-tech industries. Based on the value-added decomposition method of total export, this paper improves the RCA index, completes the replacement of value-added to export value, and uses value-added accounting system to reconstruct the index of explicit comparative advantage of value-added, as shown in Formula (6.1). DVA_{si} means the domestic added value of exports of i industry products in country s. WVA_i Represents the sum of domestic added value exported by the products of i industry in the world. E_{si} represents the export volume of I industry products of countrys and E_i represents the total exports of i industry products in all countries.

$$VRCA_{ir} = \frac{DVA_{si}/E_{si}}{WVA_i/E_i} \tag{6.1}$$

According to the RCA criterion constructed by Benedicits and Tamberi, when RCA $\geq$ 2.5, it indicates that the export of a product from a certain country has a significant comparative advantage and strong international competitiveness among similar export products from other countries, while When 0.8 $\leq$ RCA < 2.5, it indicates that a country's export of a product has a certain comparative advantage and international competitiveness of the product, and when RCA < 0.8, it indicates that the explicit competitive advantage of the product is not significant. However, VRCA index has different evaluation criteria from RCA due to different index selection. This paper sets its measurement criteria as follows. If the VRCA index of an industry in a country is lower than the average level of VRCA in other countries, the dominant advantage of the industry in the country is not obvious and the country tends to build the national value chain and accumulate its own advantages. If the VRCA index of an industry in a country is higher than the average level of VRCA in other countries the industry has certain product competitive advantages and international competitiveness and the country tends to build regional value chain and seek more profitable distribution links. What's more, the VRCA index of an industry in a country is higher than the average VRCA level of the top five countries in the statistics, indicating that the export of the product in this country has significant comparative advantage and strong international competitiveness among the similar export products of other countries. For this industry, the country tends to dominate the global value chain.

6.2.2 *Refactoring Path Selection and Criteria*

The current domestic research on high technology mainly adopts the organization for economic cooperation and development (OECD) countries five points method, which identified the manufacturing of aerospace manufacturing industry, pharmaceutical manufacturing, electronic computer and office equipment manufacturing, electronic and communication equipment manufacturing industry, medical equipment and instrumentation manufacturing as new and high technology industries.

Table 6.2 VRCA index of high-tech industry

	C9	C13	C14	C15	C27	Total average
Average level of VRCA in the world	0.75	0.73	0.65	0.66	0.67	0.69
VRCA average of the top 5 countries	2.56	2.26	2.85	2.66	2.20	2.50

Source ADB-WIOD database, WIOD2016 database and UIBE GVC Index

In terms of data selection, ADB launched the ADB WIOD database based on the World Input-Output Table (WIOD). The latest database released in 2018 covers the input-output data of 35 industries in 62 countries and regions from 2010 to 2017, including a total of 31 belt and Road countries. Other countries and regions in the database are represented by ROW. Considering the representativeness of data, the VRCA data in this paper is mainly used in the ADB-WIOD database. Since the ADB-WIOD database does not subdivide the pharmaceutical manufacturing industry, it is supplemented by the WIOD-2016 database C21 pharmaceutical product manufacturing industry. C9 (chemical raw materials and chemical products manufacturing), C13 (electrical equipment manufacturing), C14 (electronic and optical equipment manufacturing), C15 (transportation equipment manufacturing) and C27 (pharmaceutical product manufacturing) are combined with the ADB-WIOD database. This paper first calculates the VRCA index of five high-tech industry segments, and takes its average value as the measurement standard of international competitiveness of high-tech industry, as shown in Table 6.2.

Based on the analysis framework of value-added trade, the criteria for the path selection of high-tech industries to participate in the reconstruction of global value chain are as follows. When VRCA ≥ 2.5, it indicates that the export of a high-tech industry of a country has significant comparative advantages and strong international competitiveness in the similar export products of other countries. For this industry, country tends to lead the global value chain while independent innovation (this paper is called the path of active embedding into the global value chain). When $0.69 \leq$ VRCA < 2.5, it indicates that the high-tech industry has certain product competitive advantages and international competitiveness, and tends to develop local value chain while leading the establishment of regional value chain (referred to as the path of leading the establishment of regional value chain), seeking more profitable distribution links. When VRCA < 0.69, it indicates that the dominant competitive advantage of the product is not significant, and the country tends to absorb the global advanced technology to build the domestic value chain and accumulate its own advantages (building the national value chain path).

6.2.3 China's Different High-Tech Industries

The path selection of China's high-tech industry segments to participate in GVC reconstruction is shown in the following table. Electronic and optical equipment manufacturing, electronic equipment and chemical raw materials and chemical products manufacturing tend to dominate the regional value chain, pharmaceutical products manufacturing and transportation equipment manufacturing tend to build the national value chain (Table 6.3).

The VRCA index of China's chemical raw material and chemical products manufacturing industry(C9) is 0.92, ranking 20th among 62 countries, higher than the world average VRCA index, indicating that this industry has certain product competitive advantages and international competitiveness. However, in terms of industrial concentration, overall production technology and technological innovation, it still lags far behind the developed countries such as the United States, Europe and Japan, as well as the world's top three countries such as Switzerland, Denmark and Ireland. Although the bottom-up Reconstruction Path of China's chemical raw materials and

Table 6.3 VRCA index and path choice of China's high-tech industry

Industry	China	VRCA index top five in the world	Path choice of China
C9	0.92	Switzerland 2.74 Denmark 2.73 Ireland 2.63 Belgium 1.86 Slovenia 1.72	RVC
C13	1.13	Italy 2.53	RVC
		Germany 2.42	
		Czech 2.31	
		Austria 2.03	
		Hungary 1.90	
C14	1.71	Portugal 5.56	RVC
		South Korea 3.45	
		Japan 1.72	
		China 1.71	
		Maldives 1.68	
C15	0.62	Czech 2.84	NVC
		Mexico 2.82	
		Japan 2.70	
		Germany 2.62	
		Hungary 2.31	
C27	0.68	South Korea 2.99	NVC
		France 2.39	
		USA 2.36	
		UK 1.75	
		Japan 1.50	

Source ADB-WIOD database, WIOD 2016 database and UIBE GVC Index

chemical products manufacturing industry in the global value chain may be hindered by other countries, the industry still has certain competitiveness in some countries or regions. According to the different national industrial competitive and complementary, it can focus on the development of independent innovation and technology, tend to build regional value chain and expand the power of discourse of industry in areas with strong industrial competitiveness.

China electrical equipment manufacturing (C13) VRCA index is 1.13, ranking 17th in the statistics in 63 countries, higher than the world average, which shows the industry has a certain competitive advantage and product international competitiveness. Compared with Italy, Germany and other countries still have certain gap in speeding up the formation in domestic large cycle and strengthening the reconstruction of the cycle path outside the territory. According to the guidance of national and industrial policies and the convenience of spatial geography, appropriate countries are selected for regional connection.

The manufacturing of electronic and optical products in China is the most competitive high-tech industry. VRCA index is 1.71, ranking the fourth among 62 countries in the statistics, obviously higher than the world average level, indicating that the export of electronic and optical products has a strong competitive advantage and international competitiveness of products. Taking ICT industry as an example, the strengthening of R&D and ICT infrastructure have been regarded as an important driving force for the sustained growth of the global economy. Driven by the potential of the Internet of Things (IOT), global information and communication technology (ICT) companies have taken significant steps in recent years. As global technology companies are outsourcing more and more manufacturing, assembly and innovation links to more competitive regions in China, China has transferred from relying on foreign companies to achieving higher levels of autonomy and local technology innovation. China does already have a competitive advantage in brand such as Huawei, Redmi and Lenovo. However, China still lacks important core technologies such as semiconductor and so on. Core related intellectual property semiconductor concentrates on the design of the United States, South Korea, China, Taiwan and some parts of Europe. China only involves some parts such as production, assembly and testing of products. Most of the added value of semiconductor production is captured by core enterprises headquartered outside China, which hinders the path of globalization after China's independent innovation. Although China's electronic and optical product manufacturing industry has certain competitiveness, it still does not reach the global leading level. Its future reconstruction path should focus on independent innovation and local technology, and combine mobile Internet, cloud computing, big data and Internet of Things with modern manufacturing industry to adjust the industrial chain. At the same time, we need to pay attention to wage cost, exchange rate, financial factors and adjustment mechanism, select the appropriate region for value chain restructuring.

The VRCA index of transportation equipment manufacturing is 0.62, ranking 26th among 62 countries in the statistics, slightly lower than the world average level. The apparent comparative advantage of added value of this industry is not obvious. Different from the "smile curve" of the global value chain, the distribution of the domestic value chain of the transportation equipment manufacturing industry in different production links presents an "arch curve". The domestic value chain is mainly realized through the production and manufacturing links that most dependent on the added value. The upstream and downstream industries obtain relatively little added value. China's transportation equipment manufacturing industry is still at the lower end of the global value chain division of labor, and the value chain upgrades from mid-stream processing and manufacturing to the upstream of the industrial chain. As the global competitiveness of transportation equipment manufacturing industry is not strong, on the one hand, it needs to rely on the regional advantages of East China to climb up the industrial chain at the left end of the "global smile curve". On the other hand, it needs to promote the transfer of labor-intensive industries and production and manufacturing links in the east to the central and western regions, accumulate competitive advantages and improve their ability to obtain added value and build a national value chain.

For pharmaceutical products manufacturing industry, China has emerged as an important force in its global value chain. However, although Chinese enterprises occupy an increasingly high value link in the pharmaceutical value chain, due to high cost and high failure rate, Chinese enterprises rarely participate in the research and development of new concept drugs. Most domestic companies make low-cost generics or copies of existing drugs, the US and UK still dominate the life sciences sector. As shown in the table, China's pharmaceutical product manufacturing industry has a VRCA index of 0.68, ranking 14th among 43 countries in the statistics (WIOD 2016 edition), far lower than South Korea, France, the United States, the United Kingdom and Japan and other major developed countries, without obvious competitive advantages. As the traditional R&D model of large pharmaceutical companies has entered a crisis period, and the income and competitiveness are declining, the R&D model of the global pharmaceutical industry needs to be reshaped. In this context, China's main task is to develop new concept drugs and build a national value chain. At the same time, due to China's huge population and serious aging, China needs to rely on the medical service support of foreign companies for a long time while promoting the development of its pharmaceutical industry.

6.3 The Belt and Road Initiative and China's RVC

China's "One belt, one road" initiative proposed by President Xi Jinping in 2013 is not only an important measure for China to share the global development dividends, deal with the imbalance of economic development and build a community of human destiny, but also an important opportunity to create a new global governance platform and speed up industrial upgrading. The Belt and Road Initiative has received wide attention and positive response from countries along the routes, how to build an inclusive regional value chain has become a hot topic in relevant research. The deepening and summary of relevant research can be used as a reference for better promoting the construction of inclusive regional value chain and accelerating China's industrial upgrading.

As a new growth pole of the global economy, countries along "One belt and one road" are rich in resources, accounting for 63% of the world's population, 30% of world's GDP and 40% of world's trade volume. According to McKinsey, the countries along "One belt and one road" will contribute 80% of global GDP by 2050. In addition, many countries are in the stage of rapid development and have a large demand for capital, which makes it possible for China to take advantage of its own advantages to build a "self-centered" regional value chain. In particular, in the process of transferring the terminal market to emerging economies, compared with the relatively loose product demand of developed countries, it provides an opportunity for some Chinese industries to accelerate the transformation to high-end links such as R&D and build a regional value chain.

With remarkable economic growth and huge potential for development, strong industrial complementarity and comparative advantages of China's development and a good atmosphere of integration, countries along "One belt and one road" have provided a good foundation for China to build a "self-centered" regional value chain. However, targeted and differentiated measures need to be taken by China in some countries or regions due to poor infrastructure, weak industrial supporting capacity, insufficient internationalization experience of Chinese enterprises, unstable geopolitical pattern and conflict of double standards for domestic and international investment caused by insufficient coupling between domestic value chain and global value chain.

6.3.1 "One Belt, One Road" Regional Value Chain

The construction of inclusive regional value chain will transform the single and low-end global value chain embedding into "double embedding", that is, build a "Self-centered" value chain system and deepen the global value chain embedding and obtain advanced technology to promote the rise of high-end industries. For a long time, most of the research has concentrated on the industrial upgrading of single enterprise embedment, which is also the main form of China's participation in global value chain in the early stage of reform and opening up. With the acceleration and maturity of government-led construction of various parks, "group embedding" has become the main mode for Chinese industries to participate in global value chain competition. In view of the global value chain embedding mode of Chinese industry, regional value chain construction strategies based on the "Belt and Road" can be divided into the following two categories.

The first is the leadership of a single firm. Lefilleur and Maurel (2010) pointed out that transnational investment of enterprises was conducive to the formation and expansion of local production networks, thus promoting industrial upgrading. The "one belt, one road" regional value chain led by enterprises enhanced the global market share on the one hand, such as production and marketing. On the other hand, through cross-border mergers and acquisitions in developed economies and the layout of R&D networks, the RVC accelerated the formation of high-end factors "Self-centered" pattern. Large state-owned enterprises with strong policies and high level of development should become the main force leading the regional value chain. Through the construction of exemplary and leading landmark projects along the routes, enterprises can improve their visibility and competitiveness. Small and medium-sized enterprises participate in the construction of regional value chain through accepting orders from large enterprises and supporting key products. This strategy meets the needs of industrial upgrading led by large enterprises. This strategy meets the industrial upgrading needs led by large enterprises, such as large state-owned enterprises in high-speed rail, nuclear power, UHV and other industries. By building exemplary and driving landmark projects, it has significantly improved the global resource allocation capacity and international competitiveness of the industry through the constructing of exemplary and leading landmark projects, and exploring experience for Chinese enterprises to speed up "going global". The RVC should be adapted to local conditions in the face of differentiated development environment and industrial value chain governance structure. For resource-intensive countries or regions with low development level and small domestic demand market, led by resource processing industry, the driving force supporting local economy will be transferred to developed economies. Relying on the strong OEM technology of enterprises, we should promote the development of labor-intensive industries in the form of product processing and OEM production for less developed countries or regions with abundant labor force. For rapidly urbanizing and industrializing countries or regions, China will participate in investment attraction, infrastructure construction, processing trade and other activities of the host country by building various parks in the host country to form the "depressions" effect of preferential policies and locally optimized investment environment.

The second is the embedding of industrial clusters. For some industries with long industrial chains, deepening division of labor and lack of competitive large enterprises, it is difficult to form a complete industrial chain support system under the leadership of a single enterprise and they are faced with difficulties such as high costs and great uncertainties in "going global". In this regard, with the support of differentiated development and good cooperation system, the form-industrial clusters carrying out the regional cooperation capacity, industry grafting, such as construction has a relatively perfect system of labor division, closely linked to the domestic and foreign industrial clusters, which can more effectively promote along the relevant industries to enter the market, to stimulate Chinese industry to high-end link up.

In the process of industrial cluster leading the regional value chain, on the one hand, it promotes the formation of highly developed production and technology network, finely divided supply chain and productive service system, provides a guarantee for the construction of innovation and production system and plays a pioneering and demonstration role in promoting trans-regional cooperation and the transfer of domestic superior production capacity. On the other hand, putting the headquarters in China is not only conducive to confronting various uncertain risks, but also helps to realize strategic interaction among regional value chain, domestic value chain and global innovation chain by taking advantage of talents, technological advantages of developed economies, resource shortage and huge market size of emerging economies.

No matter whether it is a single enterprise or an industrial cluster, in the process of building inclusive regional value chain and climbing the high end of global value chain, differentiated approaches should be selected for different markets. For example, a large number of exports should be made for industries with obvious domestic production advantages, and overseas factories should be set up when there is a large trade friction with the host country. When there are national policy restrictions, promote technical cooperation or establish joint ventures, so as to avoid trade friction as much as possible and support the high-end rise of Chinese industry in a wider range.

6.3.2 Regional Value Chain and Industrial Upgrading

Relying on "one belt, one road" to build a multi-temporal and multi-field open and cooperative platform is not only a requirement for China to achieve balanced development of domestic economy, to integrate into the global innovation chain and accelerate of industrial upgrading, but also an important effort to promote wider economic development and rectify unbalanced economic structure. In general, the industrial upgrading mechanism relying on the "Belt and Road" regional value chain can be summarized as follows.

6.3.2.1 Global Value Chain Versus Regional Value Chain

As a high-level cooperation mechanism in a larger scope, regional value chains promote more economies to integrate into the global industrial division network and share the global development dividend by establishing close trade and investment links along the routes. In the process of regional value chain construction, the transformation and upgrading space brought by the rapid development momentum, the regional development gap, the huge emerging market and the opportunity of new rules construction provide the foundation for Chinese industries to cultivate the competitiveness of high-end links.

From a single eastward opening to the construction of a two-way open platform, it is a process of "link to chain" to "chain to chain" to support the high-end climb of China's industry with a wider range of cooperation. The first step is to realize the upgrading of domestic industries through industrial transfer, break the stickiness effect, low-end lock and the resulting "path dependence" of traditional factors of production on the industrial structure, gather resources to engage in high-end links such as R&D, design and brand marketing and gradually transform "sweat economy" into "smart economy". The second step is to provide a basis for industrial transformation and upgrading, product quality and efficiency improvement and brand influence through the establishment of industrial parks and export processing zones along the route, or the implementation of "engineering" with "industry". In particular, diversified product demand has provided opportunities for Chinese enterprises to replace imports with their own key products, further supporting their innovative development and diversification strategies. Based on high-end link, in charge of the microscopic governance mechanism will not only help to promote the Chinese standards of "going out" in the region, but also to adapt to global economic and trade rules, China promote fairer rules formulated through standard promoting the application of emerging markets, improving product market share and so on to promote "Self-centered" global trade rules.

6.3.2.2 Leveraging RVC to Support Upwardly Mobile Global Value Chains

"One belt, one road" initiative reconstructs the global value chain, and is also a process of integrating China's industry into a global value chain with deeper and wider scope and higher level. The upsurge of regional value chain support is the result of interaction between global value chain and regional value chain. The accumulated experience of long-term participation in global value chain competition has laid a foundation for China to further expand overseas investment, optimize resource allocation, improve operation and management and gather high-end elements. The embedding of the global value chain supported by the regional value chain provides impetus for the high-end development of China's industry by gathering more high-end resources and factors in the world, increase market share and brand influence to support the high-end development of the industry through the global agglomeration effect while exporting China's rich and competitive capital and production capacity to the world, vigorously introducing world energy, resources and technology. In the process of trade with technologically advanced countries, import substitution, imitation innovation and independent innovation can be realized through learning, transforming physical technology and accepting spillovers, so as to effectively enhance product competitiveness and trade gain. The industrial competitiveness that dominates the core basis of the regional value chain comes from the accumulation of continuous innovation and technology. In this process, on the one hand, it relies on the advantages of huge domestic market and developed city clusters to gather the knowledge, energy, technology and talents needed for industrial development. On the other hand, by going out and acquiring high-quality resources, technologies and enterprises, or adopting "reverse outsourcing" and other methods, high-end global knowledge, technologies and talents can support China's industrial upgrading.

6.3.2.3 Links Between GVCS and Domestic Value Chains

As the core hub of the new division of labor system, China should play a positive role in leading the regional value chain and linking domestic, regional and global markets.

While promoting China's industrial upgrading, China should drive the development along the Belt and Road and guide the integration of multiple markets in an orderly manner.

The key to the orderly integration of multiple markets is to cultivate China's capacity to lead the regional value chain under the Belt and Road Initiative, and give full play to the mechanism of connecting the past with the next and maintaining a fair global market. Relying on the scale effect of the growing domestic market, China has established its headquarters in important node cities in China, promoted mergers and acquisitions and asset restructuring of enterprises worldwide, and cultivated multinational companies with global competitiveness and "chain owners" leading the inclusive value chain. It is of crucial importance to build an efficient and collaborative domestic value chain system, form a good division of labor and cooperation pattern in product export, collaborate on innovation and overseas M&A, and at the same time, avoid the inefficient use of resources caused by vicious competition among domestic enterprises. On the one hand, the important cities along the belt and Road will be incorporated into the new value chain system through industrial transfer. On the other hand, the advanced technology and management experience gained by China's participation in GVCS will be disseminated to a larger region along the routes through the value chain channel of close cooperation. By leveraging comparative advantages along the Belt and Road, we will strengthen regional production capacity cooperation, give full play to China's leading and spreading role, and promote the overall optimization and upgrading of industries along the Belt and Road.

6.3.3 The "Belt and Road" Regional Value Chain Construction

As a more fair and open cooperation platform, it has a positive impact on China, the leader of the value chain, and countries or regions along the "One belt, one Road", which is also the core foundation for the construction of regional value chain.

6.3.3.1 China

The construction of the Belt and Road regional value chain is an effective way to get rid of the "capture" development model and gain more dividends of global economic growth. Fully drawing on the ideas and methods of developing countries in the construction of global value chain, we will gradually expand the single eastward opening pattern of participating in global competition to a two-way and all-round open system, providing opportunities for the climb of high-end industrial links.

Based on expanding the product market, resolving domestic excess capacity, extending the industrial chain and improving the utilization efficiency of foreign production factors, China has transferred some non-core links to the areas along the line, which has not only promoted the transformation from low level, extensive to high level and intensive type, but also accelerate the improvement of independent innovation capability and the development of emerging industries and advantageous industries through reverse technology spillovers in order to provide space for the upgrading of domestic industries from a "big manufacturer" to a "strong manufacturer". What's more, the expansion of Cooperation between China and countries or regions along the Belt and Road has significantly increased China's OFDI and trade exchanges with relevant regions, which is conducive to China's adaptation to international economic and trade rules, thus laying a foundation for participating in the construction of fair and open global trade rules and realizing the docking of domestic value chain and global value chain. In other words, based on greater regional cooperation, diversified market and restructured rules, China should improve the regional value chain by undertaking research and development, design and sales links with high added value, deeply integrate into the global value chain, and effectively realize the industrial climb to a higher link.

6.3.3.2 Countries Along the Routes

China, the leader of regional value chain, is integrating the global value chain and the leading regional value chain. Through the output of competitive capacity, advanced technology and management experience, China reconstructs the global governance system with the concept of "community of destiny", and effectively drives a wider range of coordinated development.

The Belt and Road Regional value chain meets the demand of countries or regions along the routes to achieve breakthrough development with the help of external forces. From the perspective of regional development, participation in the division of regional value chain brings industrial transformation, scale economies, technological spillover and optimal allocation of production factors, which not only provides sufficient jobs for local labor, but also effectively promotes the economic and social development of countries or regions and becomes an important springboard for relevant countries or regions to meet international production standards, integrate into the global value chain division of labor and improve trade profits. From the perspective of GVC embedment, on the one hand, cooperation based on infrastructure connectivity has vigorously promoted the improvement of infrastructure, diversification of development and improvement of human capital level along the routes, thus laying a foundation for host countries to integrate into GVCS. On the other hand, by changing the demand drive of developed countries to that of developing countries, enterprises along the routes will be provided with a rapidly responding demand market, which is conducive to achieving economies of scale, reducing production costs and deepening the embeddedness of global value chain.

6.3.4 *The Model of Regional Value Chain Cooperative Construction*

6.3.4.1 A "New Model of Wild Geese" Led by China

In the "Belt and Road" region, China, as a leader in innovation, technology, capital, demand and other aspects, should actively establish a China-led regional division of labor system in equipment manufacturing industry and build a "new flying geese mode". The main characteristics of this mode of division of labor are the following four aspects.

First, the region forms China as wild goose head, Russia, India, the Czech Republic, Slovakia, the equipment manufacturing industry level of development of countries along as the main body, other developing countries at relatively low development level such as central Asia, South Asia, southeast Asia, central and eastern Europe, Middle East, Mongolia Russia region as the industry division of goose tail. Among them, China is mainly responsible for upstream design and development, production and manufacturing of core components and downstream marketing services. The countries as goose body can take advantage of its own industrial base to develop capital and technology intensive industries, while countries at goose tail mainly undertakes the processing of intermediate products outsourced by upstream country. Second, in the new mode, there is an obvious industrial ladder difference between China as the leader and the developing countries along the belt and road (especially in the C17, C19 and C20 industries). The capacity and technology of equipment manufacturing industry are advanced and applicable to the countries along the belt and road. Third, China has accumulated a large amount of capital power for overseas investment, creating conditions for solving the capital problem in the process of value chain construction. China's infrastructure, such as the Asian infrastructure investment bank, BRICs Development Bank and the Silk Road fund, can help regional infrastructure construction and industrial project financing. It can also drive the strength of other financial institutions and private capital to provide strong financial support for the construction of the "one belt and one road" regional value chain. Fourth, the new mode of labor division is more stable and sustainable. One is reflected in the demand of countries along the belt and Road for different levels of industrial development, and China's market size advantage can provide a huge reselling export market from low-end to high-end for regional value chain cooperation, and reduce the impact of external market demand fluctuations and the adverse impact of powerful countries' sanctions on regional value chain. Another is building a regional value chain to help China transfer industries or production links with low added value and overcapacity to the outside world, release elements such as land, capital and labor into effective supply, realize the structural upgrading of domestic industries and change the development strategy from "imitation and catch-up" to "innovation". As the leader of the new model China's ability to catch up with more advanced countries and its traction on the wild goose array will be continuously and steadily enhanced.

6.3.4.2 Multi-level and Multi-fulcrum System of Labor Division

According to the geographical location, the "one belt one road" can be Central Asia, Western Asia, Middle East, South Asia, Southeast Asia, central and Eastern Europe and Mongolia and Russia. There are many countries with small economic volume and slow industrialization process combined with the geographical nearby areas (ASEAN and South Asia) already have a preliminary cooperation, which is difficult to directly assume a specific link in the value chain. Therefore, the value chain should be first constructed in a small range within each region according to geographical location and regional cooperation system. The small regional value chain should be taken as a whole to strengthen its connection with China and other regions and be embedded in the "Belt and Road" regional value chain.

Taking the value chain layout of C20 (manufacturing of cars, trailers and semi-trailers) along the route as an example, several small regional value chains have their own regional fulcrum and development focus. In The China-Central and Eastern Europe region, Poland, Hungary, Slovakia and other countries have relatively mature automobile industry production chains. The Czech Republic, in particular, is one of the world's best investment destinations for auto parts industry, with dense and complete auto parts industry chain and design technology R&D centers. Meanwhile, CEE countries are closely connected with the European market, and a large number of cars and auto parts are exported to The European market outside the region, such as Germany, Italy and the UK. Therefore, it is possible to make key investments in The Czech Republic, focus on technology research and development and production of middle and high-end accessories in the region, actively carry out the docking of technical standards and the improvement of market mechanism, and promote the construction of regional value chain as soon as possible. The economic level and industrial development of countries included in China-South Asia region are very different so that the level of automobile manufacturing industry and automobile market demand is also different. As a big country, South India has developed auto industry. Different from the high proportion of foreign investment in automobile manufacturing in Central and Eastern Europe, India has two large local brands, TataMotors and Ma-Hindra, whose domestic automobile industry chain is relatively complete, and the parts production technology and comprehensive customized services are relatively high. Other countries in the region, such as Pakistan and Sri Lanka, do not have a complete auto manufacturing industry. China can take India as the main investment country and enter the market through a joint venture with domestic brands. China can also make full use of metal rich raw materials with low labor cost, strengthen the development of high-end custom components and vehicle production and vigorously promote the production of low-end vehicles to meet the huge market demand in the region. For China-Southeast Asia, although countries greatly differ in development, their overall economic growth is fast and the market integration degree is high. Indonesia, as a country with a large population, has an expanding automobile industry and market scale. China can first take Indonesia as the main investment point, make use of its domestic diversified technical advantages and perfect supply chain system, as well as the abundant and low-cost labor force of other Southeast Asian countries. What's more, China can carry out a relatively complete value chain layout in the region through investment and plant construction, strengthen the construction of relevant industrial infrastructure, develop products with very different prices and make up for the shortage of industrial system in the region and develop various gradient markets in the region.

6.3.4.3 A Pivotal Role in Connecting GVC and RVC

Building one belt, one road regional value chain can change the unbalanced competition mode between the countries along the coast and the developed countries in the equipment manufacturing industry, which will link the chain competition with the chain. In this transformation, China needs to act as a connecting hub between the two value chains. On the one hand, it needs to ensure the relatively complete and independent economic operation within the regional value chain. On the other hand, it needs to promote RVC to enter the global economic operation system.

One is to connect GVC and RVC through stepped transfer of technology and production labor division. For the GVC of developed countries, China's large multinational enterprises should actively participate in the GVC led by developed countries, undertake high-quality industrial transfer from developed countries, improve the ability to learn, absorb and digest core technologies, acquire new technologies through reverse technology spillover effect, realize the catch-up strategy, and promote technological progress within the industry. For the "Belt and Road" regional value chain, China will gradually realize the transfer and grafting of the industrial chain to the region, and build a regional labor division system led by Chinese enterprises, mainly supported by connectivity construction. Relying on the construction of overseas industrial parks, we will invest and build factories overseas in areas with backward equipment manufacturing industry, transfer the middle reaches of the industrial chain, such as assembly and cost oriented processing links, or transfer the traditional equipment manufacturing links with low added value and excess capacity, and gradually form a regional labor division system dominated by Chinese enterprises. At the same time, some intermediate products, key parts and other high value-added production links will be retained in China, to help the domestic equipment manufacturing industry gradually shift to the upstream of the industrial chain of research and development and design, and develop high-end equipment manufacturing industry.

The second is to connect GVC and RVC through marketing network. As a hub country, Chinese enterprises can, on the one hand, make use of the experience accumulated in cooperation with developed countries to gradually set up brand chains in developed countries, cultivate local agents and distributors, or acquire overseas marketing channels to enhance the popularity of brands and products, so as to establish international marketing networks in European and American markets. On the other hand, products produced by the regional value chain can also be exported to developed countries in Europe and America and other countries outside the region by taking advantage of China's international marketing network and product sales channels. China acts as an intermediary for these products to enter the European and

American markets, thus helping the manufacturing links of emerging developing countries or regions along the "one belt, one road" to establish global competitive advantage.

The third is to make good use of the two-way demand relationship between GVC and RVC. In the process of development, RVC should avoid the risk of the chain being "captured" at the lower end of the existing global value chain. As a hub, China should enhance the power of discourse of the countries in the RVC of the Belt and Road in rule-making and product pricing in the world market and realize the balance and supplement of the RVC to the existing global value chain system based on its economic and trade influence.

6.4 Development of Photovoltaic Industry in the Context of RVC

With the development and integration of science and technology in various fields, emerging industries are developing rapidly around the world. In the twenty-first century, the natural science, information, life, energy and materials disciplines have opened up new technological frontiers. As an emerging industry, the current development of strategic emerging industries is at the front end of the industrial life cycle. In order to promote industrial development and maturity as soon as possible until it becomes a leading industry, understanding the endogenous growth mechanism of industrial development has significant theoretical and policy significance, which is also one of the hotspots of current domestic research.

6.4.1 The Development of Photovoltaic Industry

As a clean and renewable energy, solar photovoltaic plays an increasingly important role in the world energy structure. According to data from the World Energy Yearbook, in 2018, photovoltaic accounted for only 2.2% of the global electricity generation, but the momentum of development is rapid, with an increase of nearly 10 times compared with 2010. The proportion of electricity generation in APEC countries has increased by 10 times, and the proportion of renewable resources in the world has increased by 10 times. From 2010 to 2018, the annual average growth rate of photovoltaic power generation worldwide reached 42.8%, far outpacing the growth rate of other energy sources.

Photovoltaic industry chain includes silicon material, ingot (pull rod), slice, battery, battery module and application system six links. Generally speaking, the upstream is the silicon material related link, the middle is the cell and battery module link and the downstream is the application system link (photovoltaic power generation). The development of China's photovoltaic industry starts from the middle reaches of the industrial chain, rapidly catches up with developed countries in the upstream and realizes the market transition from pure export to domestic demand-oriented in the downstream photovoltaic power generation, which is a successful combination of energy policy and industrial policy. In terms of energy supply, according to the China Electricity Council, China's cumulative installed photovoltaic capacity in 2018 was 174.63 gigawatts, accounting for 9.2% of the country's installed power generation capacity. In 2018, it generated 177.5 billion kilowatt-hours of electricity, accounting for 2.5% of China's total. For industrial development, by the end of 2018, there were more than 2,000 large-scale enterprises in the upper, middle and lower reaches of the photovoltaic industry, with more than 2 million employees. Distributed photovoltaic, especially the development of household rooftop photovoltaic, has pulled more than 10,000 small and micro enterprises to become household photovoltaic distributors, providing hundreds of thousands of jobs for rural areas. Photovoltaic products have been exported to more than 200 countries and regions, earning a total foreign exchange of 130 billion US dollars. In terms of ecological and environmental protection, by 2018 China's cumulative photovoltaic power generation was 256.5 billion kwh, which was equivalent to reducing coal consumption by 80 million tons of standard coal and reducing greenhouse gas emissions by 210 million tons.

The development of China's photovoltaic industry in the past few years has been generally successful. The main reason is that the industry complies with China's endowment comparative advantages at all stages of development. Under the guidance of the government's energy policy and industrial policy, it has been rapidly transformed into the competitive advantage of enterprises. It has been growing in the market competition, moving from domestic to international, and the emergence of "leading" large enterprises leading the international market, which is result of the joint role of "effective market" and "promising government".

6.4.2 The Main Challenges in Photovoltaic Industry

In the whole solar photovoltaic cell industry chain in China, polysilicon and battery manufacturing have been transformed from catch-up to leading industries, while some labor-intensive industries of module manufacturing have gradually become progressive industries with the continuous improvement of Labor costs in China. China's photovoltaic industry still faces some difficulties and challenges in international competition.

The first is overcapacity. Specifically, the utilization rate of each link of the industrial chain is generally low. As can be seen from Table 6.4, the problem of overcapacity in the upper reaches of the photovoltaic industry is more serious than that in the middle reaches, and China's polysilicon capacity has exceeded the global demand. In recent years, Polysilicon production capacity in China has been gradually concentrated in western regions with lower electricity prices. According to statistics from the China Photovoltaic Industry Association, polysilicon production in western China

Table 6.4 Capacity utilization of China's photovoltaic industry in 2018

Industry chain	Capacity	Production	Utilization (%)
Polysilicon (10 thousand tons)	69	25	36
Silicon wafer (GW)	218	109.2	50
Battery (GW)	113	87.2	77
Component (GW)	151	85.7	57

Source Energy Trend and China Photovoltaic Industry Association

accounted for more than 50% of total output in 2018, up from 41% in 2017. Limited by the process, the production cost of Chinese polysilicon enterprises is higher than that of foreign enterprises for a long time so that a considerable part of polysilicon production capacity is not effective capacity, whether it can be effectively released depends on the market price. With the condition of low polysilicon price, many polysilicon enterprises in China are still in deficit, which leads to the contradiction phenomenon that China's polysilicon market is the largest polysilicon importer in the world while "overcapacity".

The second is subsidy arrears. With the rapid expansion of the photovoltaic market and insufficient collection of renewable energy surcharges, the photovoltaic subsidy funding gap reached 60 billion yuan in 2018, and it was difficult for most photovoltaic power generation projects to obtain subsidies in a timely way, which increased the capital cost of enterprises throughout the industrial chain. In particular, the photovoltaic industry is dominated by private enterprises with single business and easy to lead to operating losses due to market fluctuations. According to the 2019 results released by listed photovoltaic companies, nine of the 30 major photovoltaic companies suffered losses, with a cumulative loss of more than 7.5 billion yuan. This shows that the self-generating capacity of Chinese photovoltaic enterprises was still insufficient, mainly relying on the market size and rapid development of profit. With the decrease of subsidies, some photovoltaic manufacturing enterprises without competitive advantages will face business difficulties.

For the difficulties and challenges facing the photovoltaic industry after 2019, the key solution is to play the role of "efficient market" to assist the photovoltaic manufacturing industry to go global, for the following three reasons. First, European and American countries are always threatened by the implementation of "double counterattack" on China's exported photovoltaic products. If Chinese photovoltaic enterprises establish manufacturing bases overseas, they can avoid trade barriers. Second, when Chinese photovoltaic enterprises invest in overseas construction of photovoltaic power stations, some countries require the use of photovoltaic products produced by local enterprises as a prerequisite for obtaining preferential policies for investment and construction of photovoltaic power stations. Therefore, if Chinese enterprises directly invest in local production of photovoltaic products, it will not only help to obtain local policy preferences, but also help to solve the problem of overcapacity in the upstream segment of China's photovoltaic industry. Third, preferential policies for overseas investment and low labor costs attract the photovoltaic industry to "go global". For example, the comprehensive labor cost of Chinese photovoltaic enterprises in Vietnam is only 50% of that in China, and that in Thailand is 80% of that in China. The photovoltaic industry in China has gradually lost its comparative advantage in production and has become a progressive industry.

Data released by China Photovoltaic Industry Association show that Chinese photovoltaic enterprises have accelerated the pace of going global in recent years. Figure 6.2 shows the overseas investment capacity of Chinese photovoltaic enterprises. By 2018, Chinese enterprises had set up production bases in more than 20 countries through investment, joint ventures, mergers and acquisitions, mainly in Vietnam, Thailand and Malaysia. The effective capacity of Chinese pv manufacturers overseas reached 12.2GW of cells and 18.8GW of modules, equivalent to 10.8 and 12.5% of the domestic capacity of cells and modules respectively. In addition, the "going global" industrial chain of Chinese photovoltaic enterprises is also expanding. In the early stage of development, overseas investment is mostly component manufacturers with low cost and technical threshold. Since the second half of 2015, with the operation of China's overseas component factories gradually on the right track, the focus of enterprise investment began to extend to the upstream silicon chip, battery chip and other links, and gradually formed a local supporting industry chain.

In general, from the perspective of the needs of the development, transformation and upgrading of China's photovoltaic industry, whether from the technical,

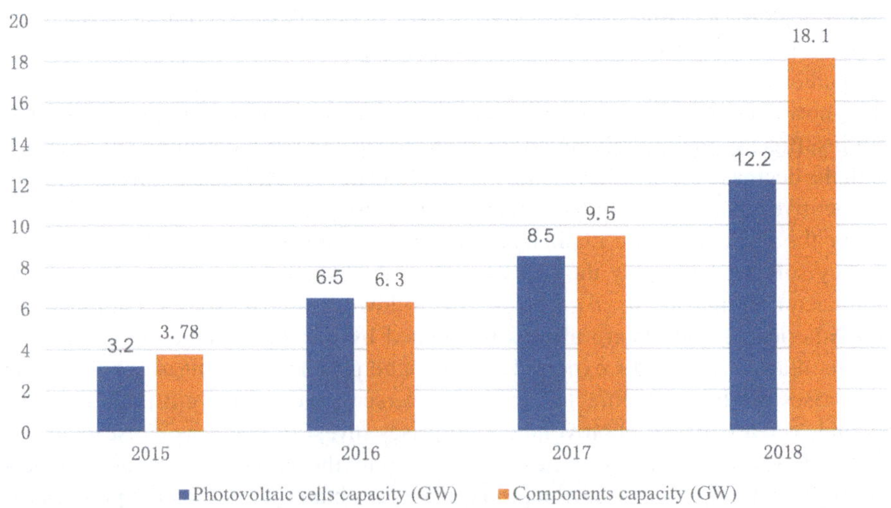

Fig. 6.2 Overseas investment capacity of photovoltaic enterprises

economic, political or strategic point of view, all support that the industry has reached the development stage of how to better transfer out. From the perspective of technology, China's transferred battery and component manufacturing capacity only accounts for less than 25% of China's total production capacity, which can be more focused on the development of high-end, high value-added advanced silicon materials and battery technology. From an economic point of view, going out of the photovoltaic industry chain can reduce production costs, expand overseas markets and avoid trade barriers, which is conducive to promoting domestic industrial upgrading. From a strategic point of view, China's photovoltaic industry going global is a practical measure to implement the Belt and Road Initiative.

6.4.3 Prospects for China's Photovoltaic Industry to Transfer to "Belt and Road" Countries

According to the International Energy Agency, there were 1.06 billion people without electricity in 2016, of which India remained the world's most populous country. As shown in Table 6.5, power supply shortages are common in developing countries along the Belt and Road. However, these countries are rich in sunlight resources, which are extremely needed and suitable for the development of photovoltaic industry. In May 2017, the National Development and Reform Commission and the National Energy Administration issued the "Vision and Actions for Energy Cooperation along the Silk Road Economic Belt and the 21st Century Maritime Silk Road", encouraging Chinese enterprises to deepen energy investment cooperation through direct investment, mergers and acquisitions, and public-private partnership (PPP). For the progressive Chinese photovoltaic industry, it is an opportunity to promote industrial transformation and upgrading and to promote the industry to the middle and high end of the global value chain.

Table 6.5 Population and proportion without electricity in some regions in 2016

Region	Population without electricity (100 million)	Proportion of population without electricity (%)
Asia	5.88	48
Developing Countries in Asia	4.39	11
Central and South America	0.17	3
Middle East	0.17	7

Source International Energy Agency

Table 6.6 Photovoltaic installed demand potential of developing countries in Asia Pacific, Africa, Central and South America (GW)

Region	2017 (actual data)	2018 (predictive data)	2019 (predictive data)
Asia-pacific (excluding China and Japan)	39	333	665
Africa	3	57	138
Middle East	1	29	91
Central Asia and Russia	–	3	4
Central and South America	5	30	55
Total	398	1589	2540

Source International Energy Agency

6.4.3.1 Potential Analysis of Photovoltaic Development in Belt and Road Countries

According to estimates by the International Energy Agency, solar photovoltaic installations in developing countries have huge growth potential. As shown in Table 6.6, it is estimated that the Asia-pacific region (countries along the belt and Road except China and Japan) will have a net increase of 294 GW of cumulative photovoltaic installed capacity from 2017 to 2030 and 332 GW from 2030 to 2040, which is the region with the greatest potential for photovoltaic development. The IEA also forecasts a net increase of 54 GW in Cumulative photovoltaic installations in Africa between 2017 and 2030 and another 81GW between 2030 and 2040. The Middle East (newly installed capacity between 2017–2030 and 2030–2040 is 28 GW and 62 GW, respectively) and Central and South America (newly installed capacity between 2017–2030 and 2030–2040 is 25 GW) are the third echelon of photovoltaic development, while central Asia and Russia are the third echelon of photovoltaic development due to the resource endowment of insufficient illumination resources. The development potential of photovoltaic is limited (only 4 GW increased from 2017 to 2040). In the Asia-pacific, Africa, Middle East, Central and South America countries, the cumulative photovoltaic installed capacity is expected to reach 902 GW between 2017 and 2040, equivalent to 34 GW of new photovoltaic installed capacity per year. If these countries no longer import photovoltaic modules from China, then China will need to transfer 34 GW of capacity overseas, which only accounts for 22.5% of China's existing çapacity (151 GW), which is fully achievable.

6.4.3.2 China's Photovoltaic Industry Chain in the "Belt and Road" Countries

In the process of industrial upgrading, developing countries have the advantage of being a latecomer and have mature industries and technologies to choose from for all industries with capital intensity from small to large. When a country in the process of economic development along the industry ascending ladder, its production scale is expanding, which in turn requires infrastructure (such as electricity supply, transportation, communications, storage hardware infrastructure and financial services and supporting policy and software infrastructure) to increase and improve. Taking Sub-Saharan Africa as an example, there are obvious disadvantages in water, electricity, communication, roads and aviation in these regions. Frequent water shortages and power outages directly lead to low production efficiency of enterprises and are also one of the major bottlenecks affecting foreign investment.

As can be seen from China's development practice, photovoltaic, as the source of power supply (the most downstream of the photovoltaic industry chain), is itself part of the infrastructure. In areas with abundant illumination and special resources, large-scale ground photovoltaic power stations can be connected to the national grid to achieve leapfrog development and become one of the main power sources. In remote areas of developing countries along the Belt and Road, small distributed photovoltaics can provide an effective solution for people without electricity. Based on the experience of photovoltaic applications in China, an investment of about $500 can solve the pain point of one household in a remote mountainous area, which is only 25% of the investment cost of the grid. Photovoltaic manufacturing, however, most developing countries at present, the application of photovoltaic entirely dependent on imports, coupled with the factors such as market size, financing costs, cost of photovoltaic in these countries (in terms of the average unit cost of whole life cycle) is 2–3 times higher than in mature markets, restrict the development of photovoltaic in these countries.

In low-and middle-income countries, if the population approaches or exceeds 100 million (such as India, Pakistan, Bangladesh, Indonesia, Philippines, Egypt, Nigeria, etc.), the photovoltaic market has a size of over GW. If the population size reaches around 50 million in middle- and high-income countries, (such as Algeria, Argentina, Brazil, Kazakhstan, Sri Lanka, South Africa, etc.), the photovoltaic application market will also reach the GW level. These countries have a considerable manufacturing base, supply chains, research and development capabilities and human resources, so they are in a position to undertake the photovoltaic module manufacturing industry and part of the mature battery manufacturing industry. Their capacity can not only meet their own photovoltaic needs (utility-scale large-scale surface photovoltaic, distributed photovoltaic), but also export photovoltaic products to other countries (including European and American markets), which are the main targets for China's photovoltaic industry to be transferred.

For other middle-, high-and low-income countries with a population of tens of millions (such as Azerbaijan, Uzbekistan, Angola and other countries), although they have a certain industrial base and human resources, the photovoltaic market is between 100 MW and 1GW, which can undertake the simple installation of photovoltaic modules. It is possible to export some products to other countries, especially small neighboring countries, while meeting the domestic demand (ground photovoltaic, roof photovoltaic, rural micro-grid, etc.).

For small countries in low-income countries (such as the Pacific Ocean, the Indian Ocean-island), photovoltaic market smaller (less than 100 mw market), no industrial base and human resources, still need to import all the photovoltaic products, power generation mode is mainly distributed (e.g., photovoltaic roof), can develop with the photovoltaic application installation, maintenance, and service business.

China's photovoltaic industry going global is not only targeted at a country or region, but also needs to consider the overall market size, industrial base and human resources of developing countries, which requires overall planning. No matter at what stage of development and with the size of China's domestic market, a country needs to cultivate professionals in planning, business model, marketing, installation and maintenance related to its own photovoltaic application market, which will also be an important part of China's "Belt and Road" technical services and intellectual output.

References

Diakantoni, Antonia, Hubert Escaith, Michael Roberts and Thomas Verbeet. 2017. Accumulating Trade Costs and Competitiveness in Global Value Chains. *World Trade Organization (WTO), Economic Research and Statistics Division.*

Lefilleur, J., and Mathilde Maurel. 2010. Inter- and Intra-industry Linkages as a Determinant of FDI in Central and Eastern Europe. *Economic Systems* 34 (3): 309–330.

Chapter 7
Challenges Faced by China's High-Tech Industry

At present, there is a clear trend of a long-term game between China and the United States around scientific and technological innovation and the development of high-tech industries. Therefore, it is necessary to make a deep study on the long-term strategic choice of American scientific and technological attack at the strategic level. This chapter is devoted to researching the current situation of China's information industry and the increasingly difficult operation of related information companies in the US market, taking Huawei as the main case, and focusing on the analysis of US restrictions on Huawei. Through the interpretation of the case, we find that although in the short term, the United States has taken a new supply chain strike against Huawei, in the long term. The restrictions imposed by the United States are a kind of market compression coordinated by technological restrictions. Huawei has been squeezed in overseas markets by a series of restrictions in the US, with market share and shipments falling. In the future, how China's information industry will face the technological competition of the United States and how to continue to gain a place in the domestic and foreign markets, it also gives practical countermeasures and suggestions at the end of this chapter.

7.1 China's Information Industry

A new round of global technological revolution and industrial transformation is poised to take off. Using information and communication technology to promote industrial upgrading and build a new driving force for economic growth has become a new trend in current economic development. The Internet and information technology are the most significant trends in the development of human society in recent decades. Information industry has become the representative of the most advanced productive forces in our life. While enabling and transforming our era, information technology has also brought new issues and challenges to human society. Network security, information security, data protection and utilization, these new things are troubling

Y. Jiang and J. Gu, *Technology and Industrial Transformation of China*,
https://doi.org/10.1007/978-981-19-7458-8_7

relevant enterprises and ordinary users, which are also gradually rising to become the focus of international politics. Worldwide, network, information and computer security, as a new branch of non-traditional security, has become the top priority of national security.

7.1.1 Overview of the Development of China's Information Industry

On February 14, 1946, ENIAC, the world's first universal computer, was introduced in the United States. In December 1969, ARPANET, the prototype of the modern Internet, was brought online by the Us Defense Advanced Research Projects Agency. Half a century later, the computing power of electronic computers has increased exponentially, and the Internet is now within reach of every household. The Internet has become another important carrier of globalization after the multinational corporation, and the Internet is far more capable of further promoting the breadth and depth of globalization than the multinational corporation. The "new human" of the information age can access all legal sources of information, make business with foreigners on different continents or instantaneously communicate with friends thousands of miles away from a smartphone or personal computer. The information technology represented by the Internet has greatly shortened the distance between people, making our planet a closer whole. From economic, political, social to cultural levels, never in history have human beings been so closely connected as they are today.

In the *Encyclopedia of China*, the information industry is defined as an emerging industry engaged in the production of information technology products, the construction of information systems, and the production and service of information content products. It can be seen from this that the information industry is mainly divided into two major sectors. The former produces hardware products such as electronic equipment and network communication facilities and the latter provides software products such as information systems and information content. Therefore, when we examine the development status of the information industry in China and the United States, the appropriate way is to start from the two dimensions of hardware and software.

In 2021, the Summit of the World Internet Conference was held in Wuzhen. The theme of this year's conference was "Towards a New Era of Digital Civilization, Joining Hands to Build a Community of Shared Future in Cyberspace". *China Internet Development Report 2021* showed that the scale of China's digital economy reached 39.2 trillion yuan in 2020, accounting for 38.6% of GDP, maintaining a high growth rate of 9.7% and becoming a key driving force for stable economic growth. China ranks first in the speed and scale of 5G network construction in the world. In addition, the conference also announced the World Internet Development Index, which comprehensively quantified the Internet development of major countries and regions in the world from six major dimensions including infrastructure, innovation capability, industrial development, network application, network security and network governance. We can learn the hardware strength and software strength of each country in the information industry from the World Internet Development Index. As shown in Table 7.1, in the overall ranking, China ranks second after the United States. Major breakthroughs have been made in high-performance computing, quantum communication, artificial intelligence, cloud computing, big data and 5G, among other key technologies at the forefront of the Internet. But in terms of specific scores, there is still a significant gap between China and the United States. China's huge population base and the rapid growth of the number of Internet users have put great pressure on the construction of China's network infrastructure. At present, the network infrastructure continues to improve, but the development is still uneven. The good news is that China's existing large stock of Internet users lays the foundation for further development of the Internet. In terms of network security, threats and risks have become a difficult problem that all countries must face and solve together. Most countries elevate the issue of network security to the height of national strategy. China's security protection level is in a medium position, and there is still a lot of room for improvement.

The World Internet Development Index shows that the Internet and information technology are the most significant trend in the development of human society in this century, and information technology has become the most representative productive force nowadays. The United States and China are the "leading countries" in the global information industry. In the foreseeable future, the overseas expansion of Software and hardware companies in China's information industry is bound to become the norm. Thanks to the successful experience in the domestic market and capital support,

Table 7.1 World Internet Development Index 2021 ranking (top 15 countries)

Ranking	Country	Score
1	United States	70.17
2	China	61.7
3	Britain	57.39
4	Germany	57.3
5	Canada	57.01
6	South Korea	56.5
7	Swedish	56.42
8	France	56.28
9	Netherlands	56.18
10	Israel	55.91
11	Japan	55.71
12	Singapore	55.45
13	Finland	54.58
14	Swiss	54.33
15	Danish	52.9

Source World Internet Development Report 2021

the influence of China's information high-tech companies in the world will continue to grow. China has become an indispensable and important member in the global information industry. As Duncan Clark says, it is getting harder to escape China's gravitational pull.

7.1.2 The Chinese Government's Plan for the Information Industry

In book *The Third Wave*, social thinker Alvin Toffler divided human society into three distinct stages. The first wave, agricultural societies, began about 10,000 years ago. The second wave, industrial society, began at the end of the seventeenth century. The third wave was the information society, which began in the late 1950s. With the decline of agricultural society and industrial society, human society is gradually moving towards the information age and entering the third wave of civilization. The most critical difference between the information age, the agricultural age and the industrial age is that the main driving force for social development has risen from physical and mechanical energy to intelligence, and informatization and the Internet are the most representative intelligence today. Because of missing the opportunity about the second wave, the Eastern countries represented by China were repeatedly beaten and humiliated by the powerful industrial production capacity of the western countries. In the third wave, although China started later than the developed countries, fortunately, the timely reform and opening policy made China the opportunity to catch up and overtake the leading countries.

At present, in the field of information industry, the United States is still a well-deserved "leader", but China has rapidly risen to become a qualified opponent of the United States, and the strength gap between China and the United States in the information industry is gradually narrowing. On the one hand, it is exciting that China's information industry has made great progress in which our national system and market size have played an irreplaceable role. On the other hand, we must recognize the objective gap between China and the United States. China may be on par with the United States in application-level hardware and software, but in more core underlying hardware and software, such as computer chips and operating systems, China still has a strong dependence on American technology and products. This gap will not be eliminated in a short period of time, because the information industry is not an industry of rapid input and output. It relies on the continuous accumulation and iteration of technology and requires long-term and stable support of talents, capital and market.

In November 2012, the report to the 18th National Congress of the Communist Party of China discussed speeding up the improvement of the socialist market economy system and accelerating the transformation of the mode of economic development. It pointed out that China needs to stick to the path of new industrialization, urbanization and agricultural modernization with Chinese characteristics. China should realize the deep integration of informatization and industrialization, the benign interaction between industrialization and urbanization, the coordination between urbanization and agricultural modernization, and promote the simultaneous development of industrialization, informatization, urbanization, and agricultural modernization. In order to achieve the above goals, China needs to focus on the construction of the next-generation information infrastructure, improve the modern information technology industry system and the information security system, and promote the application of information network technology. At the enterprise level, on the one hand, it is necessary to improve the technical level and core competitiveness of large and medium-sized enterprises. On the other hand, it is necessary to provide sufficient support for the development of small and micro enterprises, especially technology-based small and micro enterprises.

In May 2015, The State Council issued the "*Made in China 2025*" document, which is seen as a specific action plan for China's manufacturing strategy. The document proposed that with the far-reaching industrial changing caused by the deep integration of new-generation information technology and manufacturing, new production methods, industrial forms, business models and economic growth points continued to emerge. The new industrialization path with Chinese characteristics should take promoting the innovative development of manufacturing industry as the theme, improving quality and efficiency as the center, accelerating the deep integration of the new generation of information technology with manufacturing industry as the main line, promoting intelligent manufacturing as the main direction, and meeting the needs of economic and social development and national defense construction for major technology and equipment as the goal. The document established by "three steps" implement the strategic goal of manufacturing power, the first step is to 2025, the main achieve industrialization and further integration of industrialization and informatization, the construction of a number of multinational companies with global influence and industrial clusters so as to realize China's status in the global industry division of labor and value chain. The second step is to achieve full industrialization by 2035, with competitive industries equipped to lead global innovation. The third step is to build a world-leading technological and industrial system by 2049.

In October 2017, the report to the 19th National Congress of the Communist Party of China was further developed on the basis of the report of the 18th National Congress. When discussing the part of implementing new development concepts and building a modern economic system, the report of the 19th National Congress of the CPC proposed that China should keep an eye on the frontiers of world science and technology and seek breakthroughs in both forward-looking basic research and leading original results. China will make generic, high-tech and disruptive technological innovation a priority in basic applied research and major national science and technology projects. What's more, we will build a national innovation system and strategic science and technology forces to serve China's development in science and technology, quality, space, cyberspace, transportation, digital China, and smart society. The report also calls for accelerating the development of advanced manufacturing, promoting the deep integration of the Internet, big data, artificial intelligence and the real economy, and fostering new growth areas and driving forces in mid- and high-end consumption, innovation-led, green and low-carbon, sharing economy, modern supply chain and human capital services.

From the report of the 18th National Congress of the CPC to "Made in China 2025" and then to the report of the 19th National Congress, it can be seen that the Chinese government's recognition and attention to the information industry has been deepening. The significance of promoting the information industry development is not only to make China stand out in the global competition of this segmented industry, but also to transform and empower other traditional industries and fields with the help of the information industry. At present, in order to achieve the goals of social productivity improvement and comprehensive national strength, strategic adjustment of economic structure, modernization of national governance system and governance capacity, modernization of national defense and military, and construction of a smart society, it is necessary to rely on the technical and equipment support provided by the information industry. China's information industry is in a period of great strategic opportunities. The development of the information industry is closely related to the realization of the "Two Centenary Goals" and the Chinese Dream of the great rejuvenation of the Chinese nation.

7.2 American Restrictions on China's Information Industry

As the gap in information industry strength between China and the US narrows year by year, it is negatively related to the fact that the scope of the US restrictive behavior has become wider and its intensity has also been on the rise. In recent years, the United States has imposed direct restrictions or sanctions on many Chinese companies, which have seriously affected the normal operations of these Chinese companies in the United States and even the entire overseas market. This section focuses on the three aspects of foreign business restrictions, restrictions on investment in the United States and comprehensive supervision of information high-tech enterprises, discuss how the restrictions of the United States on China's information industry are increasing layer by layer and the scope of the attack is getting wider and wider.

7.2.1 Restrictions on External Business

Among the most severe and well-known restrictions on Chinese companies' foreign business is the ZTE case. On March 7, 2016, the US Department of Commerce placed ZTE on the "Entity List" for violating US export control laws and regulations, and immediately imposed export restrictions on ZTE, barring US companies from selling any technology to ZTE unless it obtained special permission. The move came after

ZTE was accused of violating US sanctions against selling US-made products to Iran and North Korea. On March 7, 2017, ZTE reached a settlement with the US Department of Commerce, ZTE agreed to pay a fine of about 890 million dollars, and the US removed ZTE from the list of restricted exports. In addition, the 300 million dollars imposed by the US Department of Commerce's Bureau of Industry and Security on ZTE could be waived if ZTE fulfilled the requirements of the agreement with the US Department of Commerce's Bureau of Industry and Security within the seven-year moratorium. In terms of personnel arrangement, ZTE needed to dismiss four senior employees involved in the case and punished the other 35 employees through bonus reduction or penalties. The incident should have been settled at this point, but in March 2018 ZTE admitted that it had fired only four senior employees, that it had not sent a complaint letter to 35 other employees until February 2018, and that all but one of the 35 employees had been awarded full bonuses for 2016. Then, on April 16, the US Department of Commerce accused ZTE of making false statements in correspondence to the US government in November 2016 and July 2017 by failing to promptly deduct bonuses and issue complaint letters to certain employees involved in historical export control violations. Regarding the actions of the US government, ZTE issued a statement on April 20 saying that the sanctions imposed by the US Department of Commerce were extremely unfair and ZTE could not accept it. At the same time, ZTE stated that it had established a compliance management committee directly led by the president.

The turning point of the US ban on ZTE occurred on June 7, 2018, when the US Department of Commerce officially lifted the export ban on ZTE, allowing ZTE to purchase parts from US suppliers and resume normal production and operation. The price ZTE paid for this was to pay a $1 billion fine and a $400 million bond, restructure its management, and send an executive team from the US to ZTE to supervise the implementation. At this point, the United States banned ZTE event officially ended. Mr. Trump said he was lifting the ban on ZTE to help with ongoing trade talks with China. Wilbur Ross, US Commerce Secretary, reassured members of Congress that "as we lift the ban on ZTE, the department would remain vigilant and we closely monitored ZTE's actions to ensure compliance with all US laws and regulations". But in The US Congress, politicians opposed to Mr. Trump's decision have not been mollified. On July 12, 2018, a bipartisan group of six U.S. senators proposed reinstating the ban. In their letter to the reconciliation committee, they said, "ZTE and Huawei were both controlled by the Chinese government, and they provided the ability to spy and steal intellectual property, and thus posed a clear threat to the national security, people, and economy of the United States".

7.2.2 *Restricting Chinese Investment by the US*

The Ford government of the United States established The Committee on Foreign Investment in the United States (CFIUS) in 1975. The original purpose of establishing this institution was to learn about foreign investment and prevent foreign investment from being politicized. In recent years, CFIUS has seen an increase in the level of suspicion and the frequency of scrutiny of foreign investments and acquisitions. Between 2005 and 2007, fewer than 5% of deals submitted to the agency were formally investigated for problems. However, between 2014 and 2016, more than 42% of transactions were investigated. Between 1975 and 2011, CFIUS blocked just one transaction. But from 2012 to 2019, CFIUS blocked five transactions. CFIUS asked Chinese company Kunlun to sell Grindr among the five transactions that did not pass scrutiny.

Kunlun purchased 61.53 and 38.47% of Grindr3 in January 2016 and July 2017 respectively with Kunlun Group Co., LTD as the main investor, contributing 93 million dollars and 152 million dollars respectively. After the completion of the equity delivery procedures in January 2018, Kunlun held a total of 100% of the shares of Grindr, realizing the overall acquisition of Grindr. According to Kunlun's 2018 financial report, as of the end of 2018, Grindr had more than 80 million users, and its revenue increased from 40 million dollars in 2015 to 86.42 million dollars in 2018. The platform's user assets grew rapidly.

In late March 2019, the Committee on Foreign Investment in the United States (CFIUS) notified Kunlun that its ownership of Grindr posed a national security risk. With pressure, Kunlun began looking for a successor to Grindr. Subsequently, Kunlun and Grindr signed the National Security Agreement with the US Foreign Investment Review Board on May 9, 2019. Kunlun must complete the sale of its entire stake in Grindr by June 30, 2020, a deadline set by the US for the spin-off. As of the effective date of this Agreement, data access restrictions shall be set. Kunlun Company, its affiliates and restricted personnel shall not access some sensitive data related to Grindr users, information systems, network connections and facilities. Grindr shall also not transmit sensitive data to or on behalf of any person or entity in China. Grindr shall also cease all operations in China and shall not hire, appoint or employ any personnel of Kunlun and its affiliates that Grindr has not previously employed. Grindr could have its headquarters located in the United States and conduct Grindr's daily business operations and management in the United States. Grindr's board of Directors consists of three CFIUS-approved persons. One person appointed by Kunlun as a director. One US citizen with mobile application technology and industry experience serves as an independent director and a US citizen with security classified levels serve as security directors.

7.2.3 US Supervision of Chinese Information Technology Enterprises

On October 8, 2019, the US Department of Commerce announced the addition of 28 Chinese organizations and enterprises (20 organizations and 8 commercial entities) to the Entity List for implementation under the Export Administration Regulations (EAR). According to the EAR, the US Department of Commerce has the authority to add individuals or organizations to its list of entities that engage in or engage in activities contrary to the national security or foreign policy interests of the United States. The individuals and organizations on the list of entities were required to obtain additional licenses for export and re-export transactions with US companies and the availability of most licenses was limited.

Eight Chinese companies affected this time are Hikvision, IFLYTEK, MEGVII, Dahua, MEIYA Pico, YITU Technology, YIXIN Technology, and Sense Time. It can be seen that the companies targeted by the US Department of Commerce's entity list were all high-tech companies in China's information industry. Among them, Hikvision and Dahua are the top two security companies in the world respectively YIXIN Technology is deeply involved in the wireless video field. IFLYTEK is a well-known listed company in intelligent voice and artificial intelligence; MEIYA Pico is a leader in the electronic data forensics industry. Sense Time, MEGVII and YITU are China's domestic computer vision unicorn companies.

From the US ban on ZTE to the CFIUS requesting Kunlun to sell Grindr to the US listing eight Chinese information high-tech companies on the Entity List, it can be seen that the US restrictions on China's information industry show a trend from point to point. The pressure is gradually escalated. At the center of the current US crackdown on China's information industry is the US's restrictions on Huawei.

7.3 The US Government's Crackdown on Huawei

7.3.1 Why Huawei?

7.3.1.1 Huawei's Rapid Development

Huawei, founded in 1987, is a provider of information and communication (ICT) infrastructure and smart terminals. It has 194,000 employees, including 80,000 in R&D. It operates in more than 17 countries and regions, serving more than 3 billion people worldwide. Huawei provides its customers with products, solutions and services in the fields of communication networks, IT, intelligent terminals and cloud services. It has 21 R&D institutes, 36 joint innovation centers and 14 research institutes (located in more than 10 cities outside China), with 2–4 R&D centers per institute. Huawei has always been known for its high investment in R&D. As of 2017, it had invested more than 5 billion dollars in research and development every year. By 2018, it had invested a staggering 70.2 billion dollars in research and development in the first ten years, including 15.3 billion dollars in 2018, ranking fifth in the world. For the company's vision, mission and strategy, Huawei describes it as follows. Bringing the digital world to every person, every home and every organization, and building a smart world connected to everything.

In addition to its main business of information and communication infrastructure, Huawei has also made continuous efforts in consumer business, cloud services,

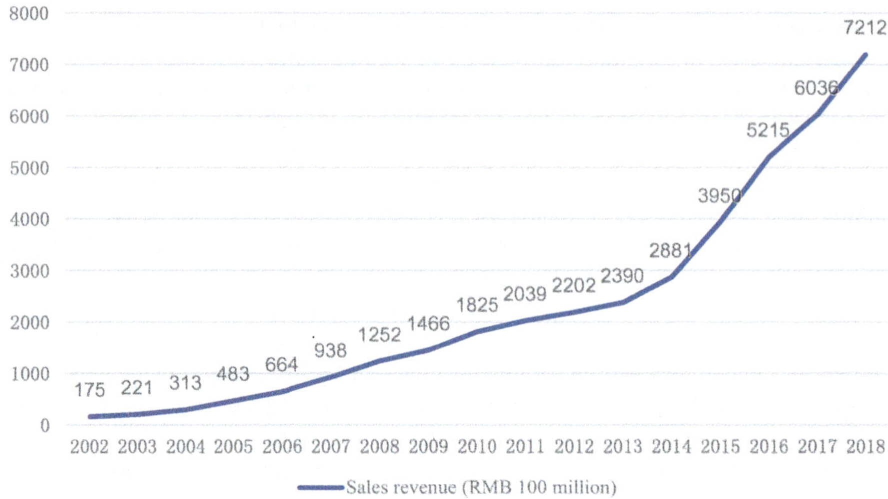

Fig. 7.1 Huawei's marketing revenue from 2002 to 2018. *Source* Huawei annual Report 2002–2018

chips and software in recent years. Among them, the sales revenue of consumer business reached 348.9 billion yuan, up 45.1% year on year, accounting for 48.4% of Huawei's sales revenue in 2018. Huawei's consumer business department is formed by the merger of the original terminal company, Internet business department, mobile phone application store, cloud computing terminal design department, and Hisilicon mobile phone chip department. Computing, application store ecological chain building, and customized development of operating systems are the development directions (Fig. 7.1).

By collecting and sorting out Huawei's financial statements over the years, the author produced Huawei's sales revenue table from 2002 to 2018, which clearly shows Huawei's growth trajectory. Huawei overtook Ericsson to become the world's largest telecom equipment maker in 2012 and surpassed Apple to become the world's second-largest smartphone maker behind Samsung Electronics in 2018. It ranks No. 72 on the 2018 Fortune Global 500 list. It can be said that Huawei is not only far behind its domestic peers, but also leads the market even compared to its foreign peers, such as Ericsson and Nokia (in fact, Ericsson, Nokia and other traditional market leaders all saw their revenues decline in the same period).

Since the 1950s, transnational corporations have achieved rapid development and become the main carriers of globalization. The extraordinary economic strength of transnational corporations naturally leads to their demands in politics. As emerging actors in international relations, their influence on international politics is increasing day by day. Transnational corporations from traditional industries such as finance, retail, manufacturing and energy reshape the world economic pattern through the networking of production, the integration of finance and the transnational monopoly of capital, and then influence the world politics and international relations. A group of new multinational companies, such as Huawei, Google, and Microsoft, were born in the information revolution. They brought new changes to world politics through information technology. What is unique about the new crop of multinationals is that their products and technologies have penetrated daily and political life to a greater extent and depth, which explains why the American government is so wary of Huawei.

7.3.1.2 What Does the US Think of Huawei?

In the 5G era of the Internet, relying on end-to-end advantages, Huawei has the technical strength and financial resources to build a strategic leadership. As the commercialization of 5G continues to accelerate, the United States has a growing sense of anxiety and urgency about competing for 5G leadership. In April 2019, The US Department of Defense's Defense Innovation Board released *the 5G Ecosystem: Risks and Opportunities for the DOD*, which focuses on the analysis of the development process of 5G, the current global competition situation and the impact and challenges of 5G technology on the US Department of Defense. The report believes that 5G, as a new generation of wireless network and technology, has a great leap in data transmission speed, capacity and delay compared with the fourth-generation network (4G). 5G will bring a host of new technologies that will re-establish public and private business standards in everything from driverless cars to smart cities, from virtual reality to combat networks. The history of the wireless age shows that the first mover countries in this field will reap billions of dollars, while also creating a large number of jobs and leading in technological innovation. At the same time, leading countries will also have the priority to develop standards and norms that other countries will have to adopt. Conversely, countries that lagged in previous wireless iterations lost the development capabilities and market potential of new-generation wireless technologies because they had to adopt the standards, technologies, and architectures of leading countries.

China, leading the field, is likely to be the leader in the early stages of 5G, which will create security risks for the U.S. Department of Defense's overseas operations, as the supply chain of its networks cannot avoid components from China. Even if the US is willing to restrict its domestic manufacturers from using products from Chinese equipment suppliers, the US market in wireless is not large enough to prevent Chinese 5G suppliers from continuing to increase market share globally, which will ultimately affect those serving the US vendors in the market cause enormous pressure. The US Department of Defense and US industrial industries will have to contend with a shift in global supply chains to China due to declining market share and competitive pressure from competing products, preventing US suppliers from investing in future 5G products. China is advancing the deployment of its first commercial 5G network, with the first Sub-6 network services scheduled to go live in 2020. This first-mover advantage is likely to drive rapid growth among China's domestic smartphone and telecom equipment vendors, as well as semiconductor and system vendors. Correspondingly, Chinese internet companies will develop services and applications based on 5G speed and low-latency performance for the domestic market. As 5G is deployed in similar frequency bands around the world, even if

China's smartphone and Internet applications and services are excluded from the US market, it is likely to remain dominant and China's development in 5G will replicate the US in the 4G era of brilliance.

Through the analysis of Huawei's own development, we can understand its leading position in the global ICT market. Through the interpretation of the report of the US Department of Defense, we explore the urgency of the US facing the challenge from China's information industry in the 5G competition. The US restrictions on Huawei are motivated by Huawei's dominance as a Chinese company in the 5G race. Huawei's size, technological strength, market share, and impact on the overall development of China's information industry make it hard for other Chinese companies in the same industry to match. The US sanctions against Huawei seriously damaged the development of Huawei and its backer, China, in the field of global network and communication facilities.

7.3.2 The US Versus Huawei

The United States began to lock down on Huawei in 2018. The U.S. strikes at the market level have limited effectiveness, while the joint allies have cut off technology supply and supply chain against Huawei. This part starts with the sudden cut off of cooperation with Huawei by the four major US operators in early 2018, and until April 15, 2021, provides a detailed review of all US actions against Huawei. According to the goals it pursues, the US crackdown measures are divided into three categories, market compression, supply chain crackdowns, and excuses for crackdowns. As shown in Table 7.2.

Table 7.2 provides a clear picture of US strikes that exhibit the following characteristics. First, in the US crackdown, the market was compressed the most frequently, accounting for nearly two-thirds of the total. After the start of the supply chain crackdown, the market compression did not stop but maintained the original strength, and both showed a clear trend of synergy. Whether it is starting from the supply chain strike measures announced by the US in May 2019 or the formal implementation of supply chain strike by the US in September 2020, the market compression of the US on Huawei has not stopped. The pattern of supply-chain crackdowns and market contractions has not changed since Mr. Biden took office.

Second, the United States' measures on market compression were very consistent, showing strong consistency, stability and continuity in the whole strike process, which indicated that the United States was trying to seek long-term effects on market compression, which was likely to be a key part of the long-term strategy of the United States. Compared with the repeated suspension of supply chain crackdowns and the wait for formal implementation, the United States had a clear target and a steadily expanding scope of its actions on market compression, and always implemented the basic principle of "immediate effect" in its actions, showing a very strong consistency. This feature has not changed since the US formally implemented supply chain crackdowns in September 2020.

Table 7.2 Events of the US crackdown on Huawei

Date	Events	Nature of event
2018 January	AT&T and Verizon of the four major US operators have terminated their smartphone sales cooperation with Huawei	Market compression
2018 February 13	FBI Director Chris Wray warned the public not to buy Huawei phones	Market compression
2018 March 22	Huawei phones lost support from retailer Best Buy	Market compression
2018 May 2	US Department of Defense banned sale of Huawei and ZTE phones on US military bases	Market compression
2018 June 20	US lawmakers asked Google to stop cooperating with Huawei	Market compression
2018 August 23	US's ally Australia banned Huawei from participating in 5G network construction	Market compression
2018 December 1	The United States instructed a third country to arrest Huawei's CFO Meng Wanzhou	Excuses for crackdowns
2018 December 5	US's ally UK operator excluded Huawei 's 4G equipment and would not purchase Huawei's 5G core network equipment	Market compression
2018 December 10	US's ally Japanese government effectively excluded Huawei and ZTE from purchase list	Market compression
2019 January 11	A Huawei employee in Poland was arrested on suspicion of spying	Excuses for crackdowns
2019 January 28	The United States has brought 23 indictments against Huawei, citing alleged theft of trade secrets and fraud. FBI raids Huawei lab in San Diego, California	Excuses for crackdowns
2019 February 27	The US delegation vigorously lobbied European governments and private companies to discourage the adoption of Huawei's 5G mobile communication technology	Market compression
2019 March 1	US warned Philippines against using Huawei 5G equipment	Market compression
2019 March 12	US warned Germany to ban Huawei or limit intelligence sharing with Germany	Market compression
2019 March 28	US's ally UK regulator has claimed that Huawei's products "significantly increase risk"	Excuses for crackdowns
2019 April 20	The CIA claimed that Huawei was funded by Chinese state security	Excuses for crackdowns

(continued)

Table 7.2 (continued)

Date	Events	Nature of event
2019 May 15	Trump signed a "state of emergency" executive order, authorizing the Commerce Department to add Huawei to the "entity list", requiring Huawei to cut off the supply of products and services that account for more than 25% of US technology	Supply chain crackdowns
2019 May 20	Google removed Huawei phones from Android upgrade list	Market compression
2019 May 23	US accused Huawei of lying about relations with Chinese government	Excuses for crackdowns
2019 June 24	FCC commissioner called for Huawei to be disconnected from US network	Market compression
2019 July 16	US Senate blacklisted Huawei on 5G legislation	Market compression
2019 November 7	Trump administration tech chief slammed countries for "opening arms" to Huawei	Market compression
2019 November 22	FCC banned carriers from using federal subsidy funds to buy Huawei and ZTE equipment	Market compression
2019 December 17	Telefonica significantly reduced equipment usage for Huawei's 5G core network	Market compression
2019 December 23	The US Department of Commerce planned to lower the US technology ratio requirement for cutting off Huawei's products and services from never more than 25% to no more than 10%, fully blocking TSMC and other companies from supplying Huawei	Supply chain crackdowns
2020 January 13	US asked UK to block Huawei from its 5G network	Market compression
2020 February 11	US government claimed to 'discover Huawei's global backdoor access to mobile networks'	Excuses for crackdowns
2020 February 13	US Justice Department charged Huawei with racketeering, theft of trade secrets	Excuses for crackdowns
2020 February 28	U.S. Senate passed bill banning government purchases of Huawei equipment	Market compression
2020 May 15	The U.S. Department of Commerce required any company that supplies Huawei with semiconductor products containing US technology to obtain an export license from the US government	Supply chain crackdowns
2020 June	US proposed "clean network plan"	Excuses for crackdowns
2020 July 29	US threatened Brazil to abandon Huawei	Market compression
2020 September 15	US trade ban on Huawei officially took effect	Supply chain crackdowns

(continued)

Table 7.2 (continued)

Date	Events	Nature of event
2020 October 14	The US announced that more than 25 EU and NATO countries have joined the "Clean Network Initiative". North Macedonia, Bulgaria, and Kosovo joined the US "Clean Network Program"	Market compression
2020 November 10	Brazil Supported US "Clean Network Initiative" Proposal on "Transparent 5G Technology"	Market compression
2020 December 21	U.S. Congress approved $1.9 billion in funding to help U.S. telecom operators ditch Huawei equipment	Market compression
2020 December 23	Ukraine has joined the US "Clean Network Program"	Market compression
2021 March 12	The Biden administration revised the relevant license to supply Huawei, further restricting the supply of Huawei's 5G products	Supply chain crackdowns
2021 April 7	US Commerce Secretary confirmed Huawei would remain on Entity List	Supply chain crackdowns
2021 April 15	Romania banned Huawei from participating in the country's 5G network development	Market compression

Source "AT & T Backs off Deal to Sell Smartphones from China's Huawei", Wall Street Journal, January 8, 2018, https://www.wsj.com/articles/at-t-backs-off-deal-to-sel1-smartphones-from-china's-huawei-1515443153?mg=prod/accounts-wsj

"Six Top U S Intelligence Chiefs Caution against Buying Huawei Phones", CNBC, February 13, 2018, https://www.cnbc.com/2018/02/13/chinas-hauwei-top>-us-intelligence-chiefs-caution-americans-away.html

So far, the US clampdown on Huawei has paid off. Huawei's global market share for 5G equipment fell fast. In 2018, Huawei was far ahead in 5G contracts and occupied an overwhelming market share due to its huge technological and cost advantages over other competitors. However, since 2019, after the continuous actions of the United States, Huawei has rapidly lost its overwhelming advantage in market share, while Ericsson and Nokia began to reverse the trend surprisingly without changing their technological disadvantages, and overtook Huawei successively.

Figure 7.2 shows the relative changes in the number of global 5G contracts between Huawei and its competitors, while Fig. 7.3 more clearly shows the sharp decline of Huawei's global market share from 2018 to the present. Based on the number of 5G equipment contracts announced between 2018 and 2020, Huawei's market share has fallen from more than 70% to less than a third.

Similar to the global market for 5G equipment, Huawei's smartphone market share overseas has experienced a rapid and sustained decline since the end of 2018. In the past two years, Huawei's overseas smartphone market share has declined rapidly (see Fig. 7.4).

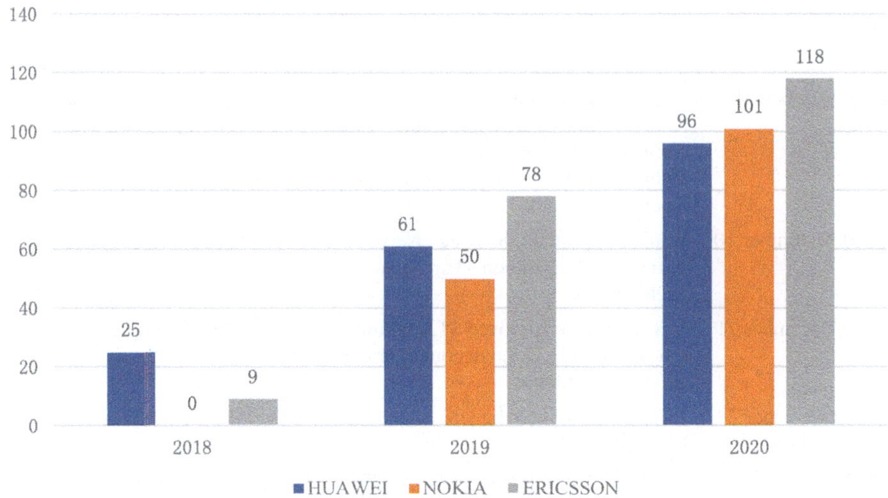

Fig. 7.2 Annual variation trend of global 5G contract quantity between Huawei and its competitors. *Source* https://www.ericsson.com/en/5g/contracts, https://www.nokia.com/networks/5g/5g-contracts/, https://www.huawei.com/en/news/2021/3/5g-private-network-gti-2021-award

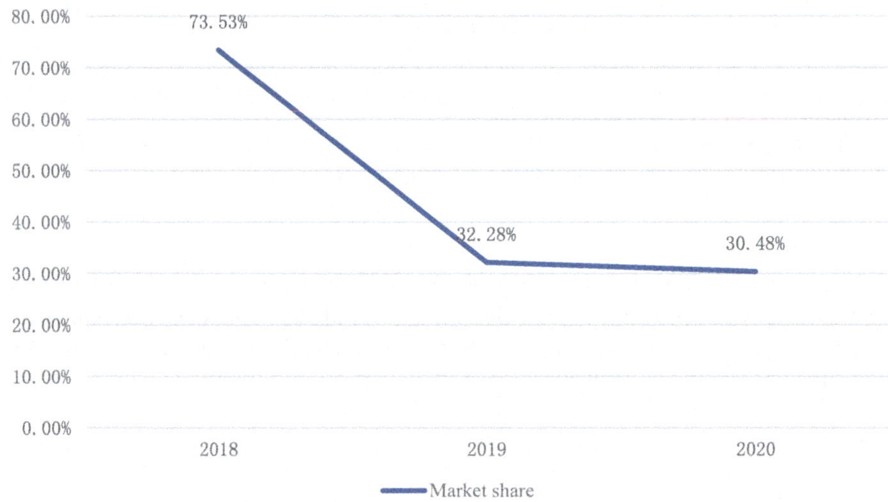

Fig. 7.3 Change in global MARKET share of Huawei 5G equipment (in terms of number of contracts). *Source* https://www.ericsson.com/en/5g/contracts, https://www.nokia.com/networks/5g/5g-contracts/, https://www.huawei.com/en/news/2021/3/5g-private-network-gti-2021-award

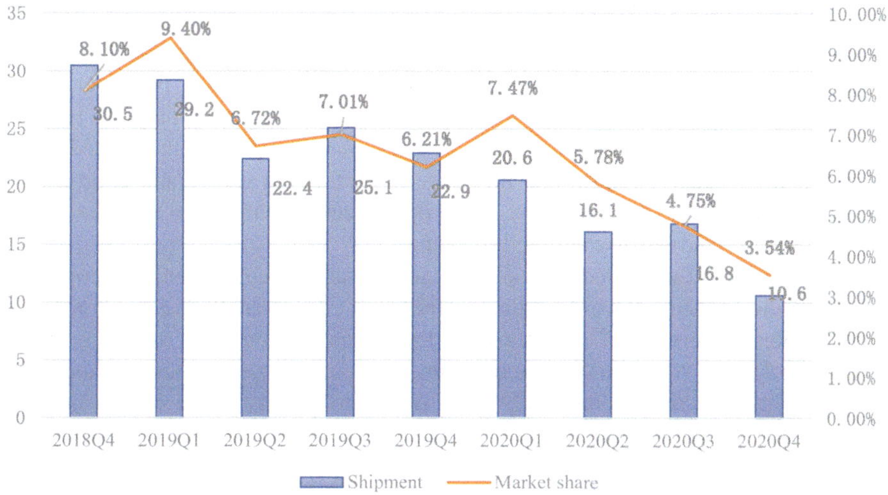

Fig. 7.4 Overseas market share and shipment trend of Huawei smart phones. *Source* https://www.idc.com/solutions/data

Judging from the specific cracking actions and realized effects, the United States has attached great importance to squeezing Huawei's market space in past actions, and the coordinated implementation of market squeezing and supply chain cracking means that the United States is likely to have adopted the strategy of "market squeezing coordinated with technological restrictions".

The United States has repeatedly raised issues in its efforts to squeeze Huawei into the market. First of all, trying to use the current law of China for conviction, for their own action a "legitimate" cloak, such action including the United States has claimed in recent years Huawei equipment safety problems. In December 2018, the United States ordered a third country to arrest Meng Wanzhou, Huawei's chief financial officer, for violating U.S. sanctions against Iran. Second, after the above actions, the US tried to establish new laws on the grounds of "national security" to provide new justifications for its continuous actions. There is a distinct characteristic of the US's behavior in raising issues. A period of intensive accusations followed by immediate action, but the United States did not provide specific evidence for these issues, and the action continued once it was initiated. This pattern of behavior reflected the nature of these issues as excuses and cover-ups.

7.3.3 Case Summary

The US crackdown on Huawei has begun to move from a short-term attack to a long-term game. Specifically, although the supply chain crackdown has achieved clear results in the short term, it has huge uncertainties in the long term. Huawei originally maintained a relationship of mutual dependence and win-win cooperation with upstream and downstream enterprises in the global industrial system, and had no intention of integrating industrial chain to build super enterprise. However, under the circumstance of technological supply failure and supply chain attack, Huawei was inevitable to abandon the illusion of US-led global free market and started to use its own technological strength and influence to establish a secure and controllable supply system. For the United States, the strikes effective in the short run were likely to create stronger adversaries in the long run.

In the long run, the United States implemented the strategy of "market compression under the coordination of technical restrictions". Its internal logic was consistent with "market compression", which was to try to push the target of attacking into a vicious circle that "the market space shrunk, the company's income declined, the R&D investment reduced, the R&D team forced to shrink, the technical strength dropped, and the market share further compressed". Finally, the strategy destroyed the whole industry and the accumulation of technical strength, maximized the blow to the opponent's scientific and technological innovation and high-tech industry development.

The containment of Huawei by the United States reflects from one side the impact of the rise of China's information industry on the leading role of the United States in making international rules. At present, the gap between China and the United States in information technology is gradually narrowing, and China has even surpassed the United States in some areas, such as the commercialization of 5G technology. This situation makes the United States worry that China will surpass the United States in technological innovation. This is because if China's innovation ability exceeds the United States, China will have more dominant power in technological standards and rules than the United States.

7.4 The US Restrictions and China's Response

7.4.1 The Future of US Restrictions on China's Information Industry

7.4.1.1 A Lasting Competitive Landscape

Today, in terms of economy, military and technology, the United States is undoubtedly still the world's leading country. With its powerful capital, technology and armed forces, the United States plays the role of the only superpower. However, emerging powers represented by the BRICS countries, such as China, Russia, India and other countries, are developing their own forces to establish new regional power centers, and threatening the hegemonic status of the United States. According to the logic of Gilpin's hegemonic stability theory, with the law of increasing marginal cost and decreasing marginal return caused by the provision of international public goods by the United States, the self-consumption of its hegemony and the decline of hegemony seem to be the inevitable outcome. The original political, economic and technological advantages of the United States will be gradually consumed and eventually be lost to the emerging challengers due to the plight of insufficient people. With the continuous rise of China's comprehensive national strength, the US anxiety and preparedness for China have also increased year by year.

During Trump's office, as he promised to voters during the election process, this US government faithfully pursued the "US first" policy, focusing on safeguarding the four core national interests of the United States, protecting people and homeland security, promoting prosperity, seeking peace with strength, and enhancing international influence. The withdrawal from the trans Pacific Partnership Agreement (TPP) agreement and the absence from the world economic forum in Davos in 2017, and the "retreat" of the United States showed that the trump administration was trying to revive the absolute hegemony of the United States in the economic and military fields in order to slow down the decline of American hegemony. Trump's foreign policy has not only led to increasing tensions in China US relations, but also led to a sharp decline in the image of the United States in the hearts of its allies. Behind the unpopular "US first" policy pursued by the trump administration, there is a basic fact that in today's international community, hegemonism is not welcome, and what the international community needs is leadership not arbitrary hegemony. The negation of hegemonism is not only a political correctness, but also an inevitable requirement for building a more fair and reasonable international order.

Affected by the atmosphere of competition between China and the United States, the probability of competition between China and the United States in the information industry will become the new normal. With the "US first" policy and the extension of trade protectionism to the information industry, the trump administration, which represents the resurgence of American Isolationism tradition, is expected to continue to adopt unilateralism to ensure the "us first" of the information industry in the future. However, when things are poor, they will change, change will be flexible, and the general rule will last for a long time. For the US restrictions on China's information industry, the Chinese government and enterprises should not only evade or counter, but also draw the United States to the track of system construction. This will not only solve the issue radically, but also enhance mutual trust and create a good atmosphere for the two countries to solve problems in other areas.

7.4.1.2 The Dispute of Technical Standards and Systems

American scholar John Ikenberry said in his monograph *After the Great War* that the state in a superior position always sought to maintain that superiority and the power that comes with it for eternity. Paradoxically, such efforts led these countries to seek to limit their own power and make it acceptable to other countries. Ishi pointedly pointed out that the international system made the exercise of power more constrained and regularized, but it also made it more durable, with the characteristics of systemic and legitimate. In the information industry, technical standards are a kind of system, and the technical strength leading the formulation of technical standards is a kind of power. The Global Positioning System (GPS) of the US used to be the only global communication system in the world, so the US basically has the right to set Global communication technology standards. Today, with the decline of the United States' technological advantages in the field of communications, its dominance in formulating technical standards is relatively shrinking so that the US feels more intense competition from China in the formulation of communication technology standards. The current conflict between China and the United States over global communication technology standards is still an interest-driven power conflict in nature, that is, the United States is worried that China's progress in network communication technology will make it impossible to continue to dominate the field. Therefore, the United States does not hesitate to take rude means of arresting people to contain Huawei. The US's containment of Huawei reflects the impact of the rise of China's information industry on the US's dominance in formulating international rules. All international rules are formulated according to the optimal level of relevant things, which means that the country with the greatest technological advantage will have the greatest right to set technical standards. This technical advantage includes two aspects, technical level advantage and price advantage. At present, the gap between China and the United States in information technology is gradually narrowing. In some areas such as 5G technology commercialization, China has even surpassed the United States. This status quo makes the United States worry that China will surpass the United States in technological innovation, because China's innovation capability surpassing that of the United States means that China will have more dominance in technical standards and rules than the United States.

In the current international community, there are more and more voices advocating the principle of cyberspace sovereignty and calling for the advancement of multilateralism. From intergovernmental organizations such as the United Nations to the international Internet community, to stakeholders such as related companies and research institutions, the formulation of rules is particularly urgent. Cyberspace norms, fundamental principles of Internet governance, and industry technical standards and norms are all to be agreed upon. With the rise of China's information industry strength, China is expected to gain more co-governance under the multi-stakeholder model. In the process of the formation of the new system, no matter whether the policy attitude of the United States towards China's information industry will change, China should play a more active role and share our wisdom and solutions on the information industry and global Internet governance with the United States and the international community.

7.4.2 China's Future Response

From the perspective of a lasting game, "supply chain ban" and "market compression with technological constraints" are two completely different strategic judgments that lead to a completely different understanding of reality and hence to completely different countermeasures. Once it is believed that the focus of the US attack is technology blockade and supply chain, it may form an optimistic judgment on the situation due to the repeated suspension of the US supply chain attack which will lead to a blind believing that the interdependence between Chinese and American industries makes it difficult for the US to achieve the attack or China's Countermeasures successfully force the US to make concessions. However, these are actually misjudgments. Only by starting from the essence of the US attack on Chinese industry and recognizing that its long-term strategy is "market compression under the coordination of technical constraints", can China be highly alert to the rapid decline of Huawei's overseas market share in the past two years and clearly realize that the phased goal of the US attack has been completed. China must take targeted and effective measures to ensure market space. Therefore, the judgment of the severity of the situation will be completely different, and the response measures to be taken will also be completely different.

For Huawei, the first thing is to insist on independent operation and decision-making and maintain openness, transparency and communication. Huawei needs to make it clear to the outside world that it is a 100% employee-owned private enterprise, and no government department or third party holds equity in the company or interferes with its operations and decisions. Although Huawei is not a listed company, it should follow the standards and practices of listed companies, maintain the tradition of releasing annual reports, and entrust independent auditors to audit the financial statements in the annual reports, so as to ensure the authenticity, integrity and independence of Huawei's business. In addition, Huawei should actively communicate with the government, media and analysts, and welcome people from all over the world to visit Huawei. Second, it is necessary to set up a special legal team to study all relevant laws and regulations of the host country, such as the export control and sanctions laws and regulations applicable by the United Nations, the United States and the EU. In addition, it is important to improve an internal compliance system through continuous investment of organization and resources. Through the policy, system, organization, process, culture, training and publicity of the company, the compliance control is implemented in all business processes and integrated into the ideology and behavior habits of every employee. Using the certainty of legal compliance to deal with the uncertainty of international politics is the best choice for Huawei and can provide it with the strongest guarantee.

Although it is urgent to deal with the technology blockade and supply chain attack by the US, we should not ignore the fundamental role of market space so as to recognize the essence of this lasting game and thus catch the opportunity due to misjudgment in dealing with the US attack. The prerequisite for Huawei to break through the technology blockade is that Huawei can maintain its own market foundation to support huge R&D investment, and at the same time use its huge market influence as a leading company to continuously radiate to the industrial chain and supply chain, promote, coordinate and integrate. Relevant technology research and development provide enterprises with market application scenarios required for technological development, and give relevant enterprises sufficient market orders to support the research and development investment required for their technological breakthroughs and finally realize the entire industrial system and supply system. On the premise that Huawei maintains its market base, the existing supply chain strangled by the US can be solved one by one with continuous efforts. However, if this premise is not realized, the effective coordination of the entire industrial system and supply system will not be realized, and it is likely to become a tree without a source to solve these technical problems. If Huawei collapses under the pressure of the American market, even a single breakthrough in one key technology may not support the recovery of the entire industrial system.

Huawei is not the first and will not be the last one to be unfairly treated for "posing a threat to US national security." For the US's suppression and restrictions on Chinese information high-tech enterprises, the Chinese government is establishing its own countermeasures. China's Ministry of Commerce revealed at a regular press conference on July 11, 2019 that China's "unreliable entity list system" was undergoing relevant procedures and would be released in the near future. The list of unreliable entities targets those entities or individuals that do not abide by market rules, deviate from the spirit of contracts, block or cut off supply to Chinese enterprises for non-commercial purposes, or seriously damage the legitimate rights and interests of Chinese enterprises. On the one hand, China's move to establish an entity list system shows that China is promoting the improvement of its legal and regulatory system in this game against the United States, which will make it more law-based in future disputes with the United States. At the same time, it also issued a warning to the United States that imposing unreasonable restrictions and suppression on Chinese companies may face equal retaliation. On the other hand, due to the lack of bilateral assessment and dispute settlement mechanisms for disputes in the information industry between China and the United States, the two sides may in pursuit of their own absolute security, drag the two sides into this kind of information security dilemma, and then pursue the realization of the "decoupling" of the two sides in the information industry, that is, independent technology and independent production in the entire industry, and finally the two countries enter into the isolation.

For the Chinese government, when domestic companies encounter US domineering behavior in overseas markets, we must offer the carrot as well as the stick. Every time a solemn protest or threat of reciprocal retaliation is issued, we should clarify its interests and express its willingness to resolve disputes with the US through consultation and negotiation. Specifically, the government needs to perform its functions in a timely and effective manner to defend the legitimate interests of domestic companies in the US market. We understand the US's concerns about national security, but it cannot be used as a reason for unreasonable and rude law enforcement to restrict the normal operation of Chinese companies in the US in an unjust manner. In addition, proper publicity campaigns through newspapers, television, and new media allow American politicians and the public to learn more facts about Chinese companies such as Huawei. Publicity should continue to convey the idea that a fair and open market environment will encourage competition, stimulate market vitality, and ultimately benefit American consumers. Second, we should foster "cordial and clean" political and business relations. While providing effective assistance to domestic companies, especially in the information industry, which involves sensitive issues such as national security, it should be careful not to deepen the distrust of the US side by excessive "force". Third, it is necessary to establish bilateral assessment and dispute settlement mechanisms with the US, so as to prevent disorderly reciprocal retaliation from triggering larger conflicts. In international trade, companies of the two countries play the leading role, but the two governments are responsible for the stage. Companies operating abroad depend on a smooth business environment provided by both governments. Fourth, we should lead the development of multi-level cooperation platforms, such as "bilateral plus multilateral" and "intergovernmental plus NGO" cooperation platforms. Information security, network security, technical standards and rules are universal issues. Security dilemmas can be easily solved using technology bilaterally or multilaterally. For example, bidirectional encryption and nearby data center placement can ensure data security.

Chapter 8
Future Transformation

As the final chapter on the industrial transformation and development of emerging economies under the conditions of globalization and informatization, this chapter focuses on the development path of China's industrial transformation with an extensive understanding of the current development status of China's manufacturing industry, where the situation is that high-tech industry is still at the low end of the global value chain. It should be pointed out in particular that the path discussed in this chapter is in accordance with relevant suggestions for curing the weakness of China's industrial development. Based on the objective understanding of China's national conditions and industrial reality, starting from the overall awareness of solving structural problems, this paper expounds the basic ideas of promoting China's industrial transformation and upgrading, and grasps the key points of future strategic development from macro and micro perspectives. At the same time, we should also clarify the role of the government and have the role better played by the government as a leading organization in this process.

8.1 Transformation of China's Manufacturing Industry

The idea of an economy's industrial transformation usually comes from its basic judgment of its own national conditions and industrial development status quo, as well as from the obstacles it meets in solving structural problems and the effective methods it seeks.

8.1.1 China as a World Economic Power

As a rapidly rising developing country, China has made great historic progress in its economy today. By the end of the first decade of the twenty-first century, China, with a

Y. Jiang and J. Gu, *Technology and Industrial Transformation of China*,
https://doi.org/10.1007/978-981-19-7458-8_8

population of 1.4 billion, had already achieved its initial goal of industrialization and generally entered the middle stage of industrialization. The rapid growth of per capita income has enabled China to complete the transformation from a low-income country to a middle-income country in just over three decades, which laid the important material foundation for the country to develop at a new historical starting point. Therefore, in the critical period of the current economic transformation, a correct review of China as the world's second largest economy is an important prerequisite for understanding its industrial transformation.

8.1.1.1 China as a Major Economic Power Worldwide Supported by the Manufacturing Industry

China is an important producer of industrial products in the world. The output value of major industrial products accounted for a large share of world industry. In 2020, manufactured industrial products accounted for 95.5% of China's total export trade. It is universally acknowledged that China has become a major manufacturer in the world, which is of great strategic significance to China's industrialization and thus its transformation into a world power. First of all, it means that China has overtaken major developed industrial countries in terms of manufacturing scale and has a strong material foundation to change its own destiny. Although in the early 1990s, China had a huge gap with industrialized developed countries such as the United States, Japan, Germany, the United Kingdom and France, with a manufacturing added value of only 145 billion dollars. But then the rapid growth of China's industry started. According to the estimates of American research institutions, China's total world manufacturing output share surpassed the United States in 2010. China has greatly improved production conditions and development environment through large-scale infrastructure construction. The most outstanding performance during this period is the enhancement of China's foreign investment utilization capacity and the explosive growth of infrastructure represented by roads, bridges, high-speed railways and airports. China is better able to cope with global crises. In 2020, when COVID-19 swept the world, China took economic development into account while preventing and controlling the pandemic, alleviating employment problems during the pandemic, minimizing economic losses and maintaining rapid economic development in the post-pandemic era. In addition, China has not forgotten to assist other countries in need, which has further enhanced China's economic and international influence. While China is coping with the crisis and achieving steady growth, the role of manufacturing in China has been changing over the past decade, from being a lagged chaser to an innovator. This transformation is also reflected in the progress of innovation of Chinese enterprises. China has produced the world's first quantum satellite communication network, and built the world's largest integrated aperture radio telescope on its own. It can be seen that the rise of China's international status and the advancement of industrialization have created extremely favorable conditions.

8.1.1.2 China's Overall Industry Development as the Second Largest Economy

China's historic progress gives us confidence in building a modern and powerful country rather than a reason for stopping or slowing down the pace of development. We need to rationally view the achievements made by the Chinese industry and China's status as the world's second largest economy, and focus on solving the structural problems in industrial development, which is of great significance to the successful realization of China's industrial transformation.

First, we need an objective estimate of China's economic power. Although China's economic aggregates have reached the second highest level in the world in terms of growth, this calculation is based on the geographical perspective of GDP, that is, the main material wealth created by the manufacturing industry that includes not only the production of Chinese enterprises, but also the value created by many foreign multinational companies. In fact, it is an important part of an economy's economic competitiveness, and it is often the most competitive value. Therefore, it is biased to judge China's economic development strength from a regional perspective. Second, we need to deeply understand the development quality of China's manufacturing industry. Although China ranks first in the world in production and sales of nearly 200 major industrial products in many industries, yet special steel that reflects a high level of production capacity still relies on imports. China's automobile production ranks in the forefront of the world, but it is greatly restricted in the design and production of engines. The number of industrial licenses approved and issued by China ranks first in the world, but the international standards formulated by China only account for 1% of the world. Third, we need to clearly understand the position of Chinese manufacturing in world manufacturing. In general, due to the lack of key technologies and core businesses, the Chinese industry is at the middle and low end of the global value chain, and the status quo of being constrained by industrialized countries in many fields has not been completely changed. China's industrial development is relatively sizable, but the weakness of core technologies makes China's industries large but not strong, lacking the autonomy of development. Based on the above, it is not difficult to find that China's manufacturing industry today is not commensurate with the status of the world's second largest economy. The industrial innovation capability and the overall development level of the manufacturing industry are the most important gaps between China and developed industrial countries. Therefore, we need to find the path to solve the problems in the process of industrial development of China, and finally realize the transcendence.

8.1.2 Solving Structural Problems

Nowadays China must take independent, innovation-driven development of its manufacturing industry as the core task, and implement a national strategy for the development of the manufacturing industry that beyond the industrial level. That is, for the survival and future development of the nation, but also for the well-being and fundamental interests of the people, the country should make efforts to improve the development quality of the secondary industry. The core strategy for breaking the current growth trap is to build a leading industrial structure in China that can promote sustainable economic development in the new era with the newest possible technological means. At the same time, based on the development of advanced technology in the manufacturing industry, the core strategy will also drive the progress of agriculture and promote the modernization of the tertiary industry such as the service industry. Ultimately, with the guidance of high technology, China can build a modern industrial system that conforms to the future development direction and integrates various industries organically. The status of a large developing country makes it necessary to implement industrial layout and structural adjustment in line with the national conditions according to the diversity of China's regional development and the multi-level characteristics of its economic development. Taking high technology as a breakthrough, China is actively committed to the development of strategic emerging industries, and strives to achieve connection with the frontier of the new industrial revolution that may soon be brought about. China should vigorously develop the equipment manufacturing industry with the support of existing industrial achievements and capabilities so as to lay a solid foundation for competition, development and national security. At the same time, China needs to take into account the needs of employment, export and regional development. In the process of continuing to give full play to labor-intensive comparative advantages, China should explore feasible ways to transform traditional industries in various regions while highlighting key points and distinguishing different target levels.

Focusing on the historical path selection of industrial development at the national level, the important work that can reflect the overall awareness includes the following points. At first, China needs to make full use of the unique advantages of the latecomers and learn all the lessons necessary to complete industrialization in a targeted method. It includes not only the correct reference to the experience of developed industrial countries, but also timely correcting for their own deviations in their understanding of future industrial development trends. In particular, we should have a scientific understanding of the relationship between the development of the manufacturing industry and the service industry. In fact, the industrial history of developed industrial countries has proved that the relationship between the two is the natural shift of the development of the manufacturing industry to promoting the growth of the service industry. The order of industrial development is not a ladder-shaped relationship with the latter replacing the former, but the embodiment of the echoing influences in the development process of China's economy.

Second, we need to clearly reveal the overall industrial development goals of the nation, and change the state of disorderly development of industries that has existed for a long time. In order to complete the industrialization of a large developing country, China must make efforts to modernize its own equipment industry and quickly reverse the passive situation in which the production of key components and technologies have long been controlled by other countries. In this process, China strives to gradually push the modern high-tech that has been mastered into industrialization according to local conditions. At the same time, according to the existing capacities and local conditions, the transformation of traditional industries will be selectively increased so as to realize the goal of transforming the latter into new industries. China should make overall planning, reasonable adjustment and gradient promotion of the three processes above so as to effectively curb the previous situation of monotonous industrial structure caused by a rush of high-tech projects in various regions.

Third, in order to effectively break all kinds of external constraints, China must put into industrial innovation in the most effective way. Although the Chinese nation has never lacked the courage and ability to innovate from ancient times to the present, there is a weakness in actual industrial innovation in our policies and systems. In order to realize the in-depth development of manufacturing industry, China needs to focus the existing human capital, especially the best talents on the main battlefield of industrial innovation with sufficient policy strength and guidance. In addition, China should also take practical measures in its systems to ensure that the advanced and practical technologies developed by scientific research institutes can actually be channeled to the industrial application departments in the first place, and make use of the collection of the respective advantages of the state and the industry to truly form a set of Chinese style innovation operation mechanism which can be adapted to the development and changes of technological economy.

Fourth, in order to achieve the grand goal of building a powerful country, we must use sufficient political wisdom to turn the actual national conditions of China, the world's largest developing country, into useful development resources. For example, the development differences around the world have been transformed into distinctive differential competition, which means transforming the imbalance of industrial development into a multi-dimensional market system of integration and complementarity. China can turn its current disadvantage of unbalanced development into the development advantage of a big country, and then seek a flexible development space for the adjustment of industrial structure and the in-depth development of industrialization, and create various favorable conditions in an all-round way. Ultimately, China can realize the sustainable development of the manufacturing industry via the appropriate historical path of industrial development, so as to realize the great rejuvenation of the Chinese nation step by step, and complete its historic transformation from a strong manufacturing country to a powerful one.

8.2 Accelerating High-Technologization

Whether a country or region can successfully complete industrial transformation depends largely on the degree of integration between traditional industries and high-tech. The realization of industry's high-technologization is of great significance to the development of traditional industries in China and to the acceleration of the optimization and upgrading processes of the economic structure.

8.2.1 The Orientation of High-Technologization

8.2.1.1 From High Energy to Green Business Model

As we all know, traditional industries are mostly labor-intensive or capital-intensive, with generally long production cycles, large investment, and high energy and material consumption. For example, in the past, the three pillar industries of the United States, steel, automobile and construction industries consumed 40% of the world's strategic materials. The high-consumption development not only fails to achieve sustainable economic development, but also leads great damage to the human living environment. The realization of green management has become a major trend in the development of traditional industries around the world. The green transformation of traditional industries mainly includes four parts. The first one is green product design. When designing the product, we should consider the end of the service life of the product can be easily disassembled, decomposition as far as possible, the components can be refurbished and reused. Second is green production process, which means cleaner production. Cleaner production emphasizes reducing energy consumption and using renewable energy. Cleaner production process should generate as little waste as possible to reduce environmental pollution. Cleaner products should reduce consumption of non-renewable resources and extend the usable life. The third one is green packaging. It refers to the use of green materials, reducing packaging costs, and not using packaging materials that are harmful to the environment. The last one is green marketing. It is in the process of product sales, to promote green concepts, to guide and strengthen consumers' green awareness.

8.2.1.2 From "Heavy, Thick and Large" to "Light, Thin and Small" Products

"Heavy and thick" is a common problem of traditional industries, especially manufacturing products, which rely on unchanging and rigid production mode on the basis of consuming a lot of energy and resources. The "thin and small" type of products is a reflection of scientific and technological progress, but also the achievement of transforming traditional industries with advanced technology. For example, with the change of people's consumption habits, large-scale radios have undoubtedly declined, but pocket-sized radios have endured for a long time. Although both have no significant difference in function, but the latter's "thin and small" characteristics can meet the requirements of people carrying, whose market demand is far beyond the former. The transformation of traditional industrial products from "heavy, thick and large" to "light, thin and small" does not necessarily require changing the function of the product, but must use high technology to achieve it. If the iron and steel industry is eager to produce alloy steel with high strength and light quality, it is impossible to use traditional smelting technology, and must rely on advanced smelting technology.

8.2.1.3 From Large-Scale Standardized Production to Personalized and Flexible Production

Most traditional industries use large-scale standardized production methods. In this way, product diversification, individuation and low cost are an insurmountable contradiction. In order to reduce costs and gain scale economies, enterprises ignore customers' personalized needs for a long time and implement single standardized product production. For the diverse needs of customers in the new economic era, this method is becoming more and more difficult to adapt to. At the same time, in order to ensure the normal operation of large-scale production, enterprises have to maintain enough inventories. It is estimated that at "normal" interest rates, inventory costs average about 25% of the cost of sales, and inventories generally exceed the value of finished goods, greatly reducing product market competitiveness. The development of high technology, especially information technology, has promoted the modernization of design technology, the precision, speed and digitization of processing and manufacturing, the flexibility and intelligence of automation technology, and the network and globalization of the whole manufacturing process. Various advanced production modes, such as CIMS, concurrent engineering, lean production, agile manufacturing, and virtual operations are also developing rapidly around the world. At present, the manufacturing automation of the traditional industry in developed countries is from the rigid connection of automatic line and automatic single machine in the 1960s, to the development of CNC, FTL (flexible automatic line), FMS (flexible manufacturing system) and IMS (intelligent manufacturing system). Manufacturing automation system is developing along the numerical control, flexible, integrated, intelligent stage. Personalized and flexible production not only overcomes the contradiction between diversification, individuation and low cost, but also reduces the production and inventory costs of enterprises. More importantly, it speeds up the rapid response to customer needs, meets the diversified and personalized needs of customers, and improves the market competitiveness of enterprises.

8.2.2 The Path for Traditional industry's High-Technologization

In order to achieve the above value goals, China can focus on the following aspects.

8.2.2.1 Applying High and New Technology to Transform Traditional Equipment

The production equipment and other technical devices of an enterprise are the main material basis of the production process. The processing and utilization of natural resources, labor productivity and product quality in production are largely determined by the level of production technology and equipment. In 1963, The Japanese automobile industry lagged behind the United States for half a century, but in the 1980s, the Japanese automobile industry surpassed the United States and ranked first in the world. An important reason is that the Japanese automobile industry adopted robots in the production line, which greatly reduced the production cost of cars. According to relevant information, it takes only 9 h to produce a car in a factory equipped with robots in Japan, while it takes 31 h for a factory without robots in the United States to produce a car. Japanese cars made with robots cost a thousand or two thousand dollars less than American cars. The mold change process, taking 8–24 h in the US factory, costs only a few minutes in the Japanese factory. There are mainly two modes to transform the backward equipment of traditional industry with high technology. One is to replace the function the other is to add automatic control system to the original equipment. The most successful case of functional substitution is to replace the traditional manual typesetting equipment with laser typesetting technology, which realizes the automatic editing and typesetting of books and newspapers by computer. In addition, the numerical control machine tool replaces the manual machine tool, and the variable frequency speed regulation motor replaces the ordinary AC and DC motor, which improves the processing accuracy of the product, reduces the material consumption and energy consumption and greatly improves the labor productivity. The control functions of traditional industries are generally completed by special control components, so the accuracy rate of product completion is low, the error rate is high, and the versatility of control equipment is poor. After realizing the automatic control of the computer, the computer controls the complex operating system according to the programming content, which saves labor and improves the efficiency.

8.2.2.2 Using High Technology to Transform the Traditional Production Technique

The transformation of traditional industry production technology is mainly divided into two aspects. First, the traditional industries themselves continue to develop in the direction of high-tech and develop new production processes. For example, new technologies such as open-end spinning, woven fabrics and non-fragmented textiles are used in the textile industry. The iron and steel industry develops continuous casting and out-of-furnace refining processes. The non-ferrous metal industry develops processes such as oxygen-enriched smelting, scintillation smelting, and large-scale aluminum smelting. The building materials industry develops cement kiln decomposition technology and flat glass float process. The railway transportation industry develops process technologies such as heavy-haul train transportation and increasing traffic density. Second, high technologies continue to penetrate and diffuse into traditional industries, transforming traditional production processes. At present, the most prominent performance in this regard is the application of computer technology to transform the production process of traditional industries. Computer aided design, computer aided manufacturing and computer aided engineering calculation systems (CAD/CAM/CAE) are widely used to speed up the automation of production processes. The United States has adopted the Computer Integrated Manufacturing System (CIMS) since the end of the 1980s, making the entire process of product development and design, planning, production, and quality assurance into a continuous computer-aided information flow so as to achieve internal logistics and information in the production system. The high integration and automation of the flow can increase the production efficiency by 40–70%, reduce the engineering cost by 3–5%, shorten the new product development cycle by 30–60%, and increase the equipment utilization rate by 2–3 times. Technological transformation of the production process of traditional industries is an important way to integrate and coordinate development between the "old economy" and the "new economy".

8.2.2.3 Using High Technology to Promote the Management Modernization of Traditional Industries

At present, the important content of traditional industry modernization management is to make full use of information network technology and implement enterprise information management. Since the 1980s, the rapid development of information network technology has brought an opportunity to the management reform of traditional industries. First of all, information network technology changes the way of enterprise organization and management. The emergence of Internet or Intranet changes the organizational structure of enterprises from the traditional vertical structure to the open horizontal structure, and the high-level decision makers can directly contact with the grassroots executive, while the traditional middle-level organization's role of uploading and sending down gradually disappears. Second, information network technology has changed the traditional transaction process. Based on the information network, enterprises use electronic data to exchange business data and carry out trading activities, which greatly simplifies the trading process and improves the economic efficiency of enterprises. Third, information network technology has changed the traditional marketing management mode. The enterprise marketing management with the information network environment is the network interactive marketing management. The role of the customer in corporate marketing has been enhanced, and buyers and sellers can interact anytime and anywhere (instead of the traditional one-way communication in corporate marketing). Finally, information network technology realizes network customized marketing management. The Internet has improved the relationship between enterprises and customers. With the increase of mutual understanding between enterprises and customers, sales information will become more customized, and electronic marketing will turn mass sales to customized sales. For example, the US General Motors Buick brand car factory provided a service system that allowed customers to design their favorite car structures in front of the computer terminal in the car dealer's showroom.

8.2.3 Strategic Choice for High-Technologization

Based on the above, we have found a general path for developed countries to realize the traditional industry's high-technologization. According to China's actual industrial development status, traditional industries are still the pillar industries of national economic development. 87% of China's GDP is created by traditional industries, and about 70% of the national fiscal revenue comes from traditional industries. Therefore, it is the main task of China's current economic development to vigorously promote the high-tech transformation and modernization of traditional industries.

8.2.3.1 Breaking and Eliminating Institutional Barriers

First, we need to deepen the reform of the enterprise system to truly make enterprises dynamic players in the market. Enterprises are full of vitality, sensitive response to market signals, and have strong desire for innovation and expansion, which are the foundation and premise of traditional industries' high-technologization. Therefore, it is necessary to take property rights reform as a breakthrough point, clarify the relationship between enterprise property rights. At the same time, we should reform the cadre selection system of state-owned enterprises and form a mechanism for the survival of the fittest for operators so as to ensure state-owned enterprise really become the market competition main body, have the constant pursuit of economies of scale, technological innovation and improve the market share of intrinsic motivation and pressure. Second, establishing a unified, perfect and orderly competition market system and improving the market competition mechanism. In market economy, technical innovation is more about corporate behavior rather than government behavior. It is necessary to speed up the transformation of government functions, resolutely break the monopoly of industries, departments and regions and form a unified and open market system nationwide. In addition, China also needs to further improve the commodity market, cultivate and standardize the production factor market and property rights market, reduce unreasonable barriers to industrial entry and exit, and enable resources to be effectively cross-industry, cross-sectoral and cross-regional in order to create a good institutional environment for optimizing the industrial organizational structure and improve the efficiency of resource allocation.

8.2.3.2 Accelerating Industrial Organization Policy Oriented by Agglomeration

The low degree of industrial concentration has seriously restricted the process of high-tech in China's traditional industries, which is manifested in the following aspects. First, the scale and speed of capital accumulation of low concentration cannot meet the investment requirements of high-tech. The low degree of industrial concentration made the production scale of most enterprises in China far below the requirements of the industrial economic scale. The production cost of enterprises remains high and many enterprises have been in a state of low profit margins and losses for a long time. Enterprises lack the ability of capital accumulation, there is no chance to seek tens or even hundreds of millions investment fund for high technology development.

Second, the low degree of industrial concentration restricts the technological progress of the industry and the innovation activities of enterprises. Empirical analysis shows that a certain degree of concentration of the industrial environment is conducive to scientific and technological progress. Most revolutionary technological innovations can only be produced through long-term professional research and development, and the required R&D funds and risks can be borne by large-scale enterprises. Although the innovation activities of small and medium-sized enterprises are also very active, most of their innovation is concentrated in the field of extended technology, mainly in the diffusion of innovation. Moreover, due to the lack of sufficient financial support, technological innovation can't quickly transfer to large-scale production. In addition, SMEs do not have the strength to guarantee the benefits of innovation. Once the innovative achievements are imitated by a large number of free, it will affect its enthusiasm for innovation and affect the technological progress of the entire industry.

Third, low concentration has led to price wars in China. Even if a price agreement is formed, it is difficult for all enterprises to abide by it for a long time. Any enterprise trying to make profit by lowering the price of incomplete information will make the price agreement aborted. Price wars in the industrial market fluctuate continuously, which weakens the research and development ability of enterprises, keeps the products in low grade and low level for a long time, and hinders the upgrading of industrial technology. Therefore, the government should implement the policy of industrial organization oriented by industrial concentration, strengthen the investigation and research on the current situation of China's industrial organization structure, formulate the minimum scale standard for the development of enterprises within the industry, and form a governmental barrier to hinder the entry of enterprises lower than the economic scale and prevent excessive competition within the industry.

8.2.3.3 Cultivating the Core Competitiveness of Enterprises

Paying attention to the core function and realizing the specialization of production is the result of social division of labor and also the requirement of the development of science and technology. With the rapid development of science and technology, product technology structure is increasingly sophisticated and complicated. It is difficult for an enterprise to undertake all production links while constantly developing all relevant technologies, so that all technologies have reached the advanced level. The simplification of production content can make the production activities of enterprises concentrated in a smaller scope, which is conducive to technological accumulation and innovation on this basis, and maintain the advanced nature of industrial technology. At present, industries, regions and enterprises in China are often self-contained, and it is difficult to form a reasonable professional labor division and cooperation system between large and small enterprises so as to achieve the benefits of professional division of labor. For example, in the machinery industry, 80% of the 120 thousand machinery industry enterprises belong to "all-round" enterprises. In the machinery industry, self-made pieces and forgings accounted for more than 80% or even 90%, while the similar foreign enterprises in the United States from the casting accounted for less than 40%, Japan accounted for about 15%. The non-specialized omnipotent production disperses the resource allocation and weakens the innovation ability of the enterprise. Therefore, the enterprise must concentrate resources as much as possible, cultivate the core competitive ability of the enterprise along the route of the enterprise's technology orientation, and maintain the leading position of the enterprise in the main business fields.

8.3 Assistance of Disruptive Technology to High-Tech Industries Transformation

In addition to grasping the transformation of the manufacturing industry from a global awareness, we also need to achieve innovation from the fundamental driving force. Innovation is the first driving force for development, which is also the strategic support for building a modern economic system. With the acceleration of a new round of scientific and technological revolution and industrial revolution, disruptive technological innovations are emerging all over the world, which will provide a historic opportunity for China to transform its economic development mode, optimize its economic structure and transform its growth momentum. In the technological innovation system, the status and role of breakthrough technological innovation are becoming more and more prominent. At the same time, China's economy has moved from a stage of rapid growth to a stage of high-quality development. Accelerating industrial transformation and replacing old drivers of growth with new ones have become an important task of high-quality economic development. In this process, disruptive technological innovation plays an irreplaceable role.

8.3.1 The Mechanism of Transformation

Industrial transformation is an organic whole formed by the upgrading of industrial chains, value chains, innovation chains and production factors, which the industry from the low value-added, low technology level state to a state of high value-added and high technology evolution, mainly reflected in the technological upgrading, product upgrading, functional upgrading and industrial upgrading, etc. Disruptive technological innovation is an important driving force for industrial transformation and upgrading, and it is also the only way for latecomer economies to get rid of technological dependence and achieve leapfrog catch-up.

8.3.1.1 Stimulating Industrial Technology Orbital Transition

From the perspective of technology track transformation, technology development and evolution present an S-shaped trajectory and move along technology track. Different from the technological track of incremental technological innovation, disruptive technological innovation often creates a new technological track and economic paradigm, which has a subversive impact on the existing technological track. At the same time of realizing the technological track transition, the new industrial technological track is further formed and is embodied in jumping from one S-shaped curve to another S-shaped curve, which is a kind of nonlinear technological leap. The application of technological leap is to use the disruptive technological innovation to develop new products, thereby improving the competitiveness of products. Technological transition realized by the technology track leap plays an important role in the process of upgrading leading industries. On the one hand, disruptive technological innovation promotes the formation of new industrial technology system and the upgrading of the entire industrial system through core technological innovation. On the other hand, the transition of industrial technology track drives the extension and expansion of industrial chain and value chain, and therefore the innovation subjects gain product, technology and market competitive advantages.

8.3.1.2 Accelerating the Expansion and Extension of Industrial Boundaries

Technological innovation is the internal driving force to promote industrial integration. The diffusion and application of technological innovation with the influence of enterprise-wide economy and market demand, makes the boundaries between different industries tend to blur, and horizontal industries to form a new type of industrial organization that competes and cooperates with each other. Disruptive technological innovation makes micro-enterprises move to the two ends of the smile curve of the value chain, forming the core competitive advantage of enterprises, thereby improving enterprise performance. Industrial convergence driven by technological innovation also changes the industrial market structure and industrial performance, and promotes the change of industrial structure and economic growth pattern at the macro level. The industrial structure and the mode of economic growth have changed. For example, driven by the new generation of information technology, China's manufacturing industry presents a service-oriented trend, while the service industry tends to be productized. In the course of economic development, the deep integration of breakthrough technological innovation and traditional industries is not only an important way to enhance industrial competitiveness, but also conducive to the transformation and upgrading of industrial structure, and will release a multiplier effect on economic growth.

8.3.1.3 Upgrading the Industrial Innovation Chain

The industrial innovation chain is composed of a series of innovative activities such as basic research, technology research and development, practical application, industrialization and marketization, which is an important part of the industrial ecosystem. Major technological innovation provides innovative knowledge and technology supply for industrial development, permeates and spreads across different industries, and continuously realizes the industrialization and marketization of technologies. Disruptive technological innovation also drives complementary innovations and upgrades, including core, common, and key technological breakthroughs, as well as expansion and upgrades in original innovation, introduction, digestion, absorption innovation, integrated innovation and collaborative innovation. At the technology industrialization and market application, the application and promotion of pioneering and disruptive technological innovation forms new products, new formats and business models, and plays an iterative and upgrading role of innovation in the middle and lower ends of the industrial chain. From the perspective of the relationship between the innovation chain and the industrial chain, they interact and depend on each other. The innovation chain is embedded in the industrial chain and is the basis for the value increment of all links of the industrial chain, especially the R&D and manufacturing links. Every link of the industrial chain may become a breakthrough of innovation, thus producing a new innovation chain.

8.3.1.4 Promoting Industrial Chain Restructuring and Value Chain Upgrading

Industrial transformation driven by technological innovation is manifested in industrial chain reconstruction and value chain upgrading, which will improve and optimize the labor division and cooperation and element combination among enterprises in the industrial chain, enhance the functional state of the industrial system, and thus promote the industry to climb to the high end of the value chain. Enterprises are the subject of technological innovation and the micro basis for industrial transformation. Disruptive technological innovation is a key factor to enhance enterprises' core competitiveness and value creation ability and lead the development of the industry. Technological innovation generates new products, new technologies or product function upgrading and improvement, which accelerates product replacement and thus promotes the upgrading of the value chain. For enterprises, disruptive technological innovation results not only enhance the added value of products, but also gain competitive advantages in basic research and development, market competitiveness and other aspects, thus realizing the upgrading of the value chain. Taking Huawei as an example, for more than ten consecutive years, the ratio of R&D investment to sales revenue has been above 10%. According to the European Commission data, Huawei ranked sixth in the world and first in China in R&D investment in 2017, surpassing Apple, Google, Microsoft and other enterprises. Huawei's growth process is a process in which products continue to move from the low-end of the industry chain to the high-end. The important reason for enterprises to step into the high-end links of the industrial chain is forming the international brand awareness and competitiveness of the enterprise by the research and development of disruptive technological innovation.

8.3.2 Transformation Versus Disruptive Technological Innovation

8.3.2.1 Accelerating the Transformation of Traditional Industries

Through structural adjustment and the application of disruptive technologies, traditional industries can not only improve production efficiency and business performance, but also shift the focus of production to industrial fields or value chain links with high product added value. In particular, technological transformation focusing on digitization, networking, and intelligence will continuously strengthen the role of advanced basic processes, core components, and industrial technology infrastructure, and even subvert traditional production and operation methods. For example, the application of 3D printing technology in the manufacturing field is a typical disruptive technological innovation, which makes the mass customization impossible to become a reality, and the new technology of additive manufacturing replaces the traditional material reduction manufacturing method. It will have a profound impact on the combination of traditional technological processes, production lines and industrial chains, and provide good technical support and development space for the intelligent transformation of the manufacturing industry. At present, 3D printing technology has been widely used in industrial manufacturing, aerospace, automotive, architecture, medicine, cultural creativity, archaeology, cultural relic restoration and other fields, with broad market prospects.

8.3.2.2 Promoting New Fields of Industrial Development

Driven by disruptive technological innovation, business models, production organization methods, and industrial forms will also undergo revolutionary changes, open up new space for industrial development, and continuously evolve new industrial forms, thus becoming new economic growth points. For example, the development of a new generation of information technologies such as the Internet of Things promotes the evolution and upgrading of network technologies such as 5G, Narrowband Internet of Things (NB-IoT), and edge computing, and promotes the development of the information technology industry and the improvement of its application level. Moreover, with the diffusion and penetration of a series of disruptive technological innovations driven by the new generation of information technology, the industry has been continuously promoted to develop new fields and achieve a leap in productivity. In recent years, the integration and development of a series of information technologies such as artificial intelligence, big data, cloud computing, and the Internet of Things technology has led to the vigorous development of industries such as intelligent networked vehicles, drones, and robots. The value of IoT technology applied in the factory environment is not only in the improvement of production efficiency, but also in the aspects of safety production, equipment maintenance, inventory optimization, which can create 1.2–3.7 trillion dollars of economic value annually.

8.3.2.3 Driving by "Technology-Market" Synergy

Mowery & Rosenberg put forward the interactive mechanism theory of "market and technological innovation" and market demand promotes enterprises to realize a virtuous circle of technological R&D innovation. In terms of technology, the most significant feature of breakthrough technology innovation is that the disruptive new technology replaces the original mainstream technology and impacts the market. The formation of enterprise technology and the maintenance of competitive advantages require continuous accumulation, investment and self-breakthrough. Otherwise they would face the ever-changing market environment and technological innovation competition. Enterprises that have achieved breakthrough technology leadership would have an innovation crisis in the next round of competition, such as Nokia mobile phones. In the era of smartphones, Kodak cameras were replaced by digital cameras and other cases. Therefore, enterprises continue to update and iterate in the research and development of breakthrough technological innovations, seek new technological breakthroughs to play a leading role in the market, and promote industrial upgrading through continuous technological upgrading in this process. At the market level, potential market demand is the internal driving force to induce enterprise innovation. Disruptive technological innovation can change the existing market pattern, obtain new markets, greatly expand the profit space of enterprises, and thus form new growth points of enterprises. Herrmann and others pointed out disruptive technologies can not only better meet the needs of users in existing markets, but also further explore the potential market to meet the needs of users at a higher level. On the whole, the two subsystems of disruptive innovation technology and market cooperate and interact with each other. Market demand promotes the birth of disruptive technology, and technology accelerates the formation of innovative products, triggering new market changes and driving the development of relevant industries.

8.3.2.4 Technology Diffusion and Penetration Among Industries

Once a new Technology was formed and recognized in product application and market demand, it would continue to spread across industries, especially Genera-Purpose Technology (GPT). Although both progressive technology innovation and disruptive technology innovation have this kind of diffusion effect, there are great differences between them and disruptive technology innovation has stronger diffusion and infiltration effect. With the expansion of market demand, the market share of disruptive technology innovation gradually increases, and even has a certain market monopoly power, which leads to changes in the market competition pattern. As the number of new entrants in new industries increases, copycat innovation, technological improvements or product updates allow breakthrough technologies to spread to the market. At the same time, breakthrough technological inventions or product applications spread to other industries, forming batch after batch of new knowledge-intensive industries. For example, the fifth-generation communication technology (5G) is recognized as a typical breakthrough technological innovation in the field of information technology, which not only promotes great changes in the upstream and downstream industries of the traditional communication industry chain, but also continuously spreads and penetrates in many industrial fields such as processing and manufacturing, energy, finance, retail and transportation.

8.3.3 New Generation of Artificial Intelligence Industry

After more than half a century of accumulation and development, artificial intelligence technology has become a strategic technology leading a new round of scientific and technological revolution and industrial transformation, which is recognized as one of the three high technologies (energy technology, space technology, artificial intelligence) in the world today. The new generation of artificial intelligence technology driven by big data is a typical case of technology-led innovation. With the development of big data, machine learning and other related technologies, AI accelerates the diffusion and penetration of production, consumption and other fields, especially in improving social labor productivity, effectively reducing labor costs, optimizing products and services, and constantly promoting the transformation and upgrading of the three industries.

8.3.3.1 Agricultural Modernization and Intelligent Development

The new generation of artificial intelligence technology has injected new vitality and new momentum into modern agriculture, which has been widely applied in all aspects of agricultural production, bringing new opportunities for agricultural development and constantly improving the automation and intelligence level of modern agriculture. On the one hand, in large-scale agricultural production such as field farming, breeding, and agricultural product harvesting, the use of automated agricultural machinery, robots and other intelligent equipment has replaced traditional manual operations, saving labor costs and improving production efficiency and quality. On the other hand, in all aspects of agricultural production, the application of intelligent technology has improved the accuracy of production decisions.

The artificial neural network, deep learning and intelligent recognition technology have great application value in monitoring analysis of soil composition and fertility, crop seed identification, irrigation water quality analysis and control, breeding, agricultural facilities, Chinese caterpillar fungus ill recognition, intelligent agricultural harvest, quality detection. In the process of agricultural production, sensing technology is used to collect data in real time, such as light, temperature, humidity, CO_2 concentration, and moisture required for crop growth, and to analyze big data intelligently so as to provide decision-making basis for production managers. Alibaba's "ET Agricultural Brain" applies artificial intelligence to fields such as pig breeding, melon and fruit planting management, and realizes functions such as agricultural digital file management, full life cycle management, intelligent agricultural analysis, and full-link traceability. The application of artificial intelligence in agriculture can realize the optimal allocation of modern agricultural resources and the integrated management of production, supply and marketing.

8.3.3.2 The Transformation of Industrial Intelligence

Industry is the most widely applied field of artificial intelligence, and the breakthrough technological innovation of artificial intelligence is pushing industry into the era of intelligence. Developed countries are competing to lay out artificial intelligence strategies to seize the strategic heights of the industrial field in the new round of economic growth, such as the Industrial Internet in the United States, Industry 4.0 in Germany, the "Super Intelligent Society 5.0" strategy in Japan, and the "modern industrial strategy" launched by the United Kingdom. The unique advantages of artificial intelligence can effectively solve the problem of insufficient labor force and high labor cost. Industrial robots, intelligent factories, unmanned factories and intelligent manufacturing reduce labor demand and greatly improve industrial production efficiency. According to Fraunhofer Research Center in Germany, AI can increase productivity by 0.8–1.4% per year. Artificial intelligence technology has also brought about changes in the industrial production mode. Traditional mass production is gradually moving towards flexible manufacturing, making it possible for mass customization to meet the personalized needs of consumers. AI technology runs through the whole production cycle of product research and development, supply chain operation and intelligent services. Among them, AI core technologies such as automatic identification, human–computer interaction, big data and machine learning have been widely used in intelligent design, mass production, remote management, health assessment and fault prediction. The application of artificial intelligence technology has created favorable conditions for improving product manufacturing quality, enhancing the international competitiveness of manufacturing industry and realizing the upgrading of value chain. The development of industrial intelligence has become the core link of revitalizing the real economy.

8.3.3.3 The Service Industry Innovation

Intelligentization of service industry is the development trend of emerging service industry driven by the new generation of information technology and is also the key path to accelerate the transformation and upgrading of service industry. It is necessary to make full use of artificial intelligence technology to vigorously improve the efficiency and service quality of the service industry. Using big data, artificial intelligence and other technologies to mine market information and study consumer preferences can provide products that meet consumers' individual needs and further develop market potential. At the same time, the construction of artificial intelligence technology-oriented supply chain system, improve service quality, improve enterprise operation efficiency. The intelligentization of the service industry drives the development of emerging service industries. At present, the development of artificial intelligence industry is developing rapidly in education, medical care, pension, financial services, smart home, cultural creativity, public services and other fields. The traditional service industry continues to develop artificial intelligence technology, and the application market has broad prospects. The development of the service robot industry is relatively intuitive, which reflects the development status of China's intelligent service industry. According to data released by the Chinese Institute of Electronics, China's service robots are developing rapidly. The market size in 2020 is 28.38 billion US dollars, a year-on-year increase of about 37.4%. It is expected that the service robot market size will exceed 60 billion yuan in 2023. In terms of "AI plus emerging service industry", we need to strengthen the development of smart logistics, smart interaction, smart travel, smart wearable devices, technology finance and other fields. As a result, artificial intelligence will generate huge economic and social benefits, and accelerate the high-quality development of the service industry.

At present, the fourth industrial revolution led by the application of a new generation of information technology is accelerating. With the gradual diffusion and penetration of technology and its deep integration with the real economy, many disruptive technological innovations have been spawned, such as unmanned driving, block chain, quantum communication, 3D printing and intelligent robots. The born of these technological innovations not only changed people's production and lifestyle, but also brought revolutionary influences in improving social labor productivity, reducing labor costs, promoting the transformation and upgrading of traditional industries, and even reshaping the global industrial competition pattern. In order to fully grasp the strategic opportunities of the fourth industrial revolution, accelerate the formation of breakthrough technological innovations, and promote the transformation and upgrading of industries, China should focus on taking the following measures.

First, we will strengthen innovation in basic research and promote disruptive technological innovation that is the accumulation of long-term basic research. Basic research has the characteristics of pioneering and strategic, and is the key field to produce major original innovation results. Through the national Science Foundation projects and major scientific and technological research programs, we will focus on supporting breakthrough technological innovation research, guiding enterprises to engage in basic research of industrial needs and establishing a national platform for technological exchange, cooperation and transformation. What's more, we need to give full play to the guiding role of financial funds. For basic research projects with great social benefits and wide influence, the government shall vigorously provide fiscal subsidies, special support and tax incentives to reduce the investment risks of enterprises.

Second, we will increase support for ground-disruptive technological innovation with industry commonalities to lay the technological foundation for industrial transformation. Global competition in technological innovation is becoming increasingly fierce, and some key technologies in China's industries are subject to human constraints and generic technology support systems can't meet the needs of industrial transformation. Therefore, we need to make overall plans, strengthening industry key technology breakthrough in common strategic layout, led by the national science and technology department to establish a common key technology development planning, with the core key technologies which restrict the development of the industry as the main direction, focus on capital, talent, innovation platform and other advantage resources, through project leading, market-oriented operation, innovation efficiency promotion.

Third, we need to transform and upgrade traditional industries by using next-generation information technology and industrial integration. The new generation of information technology, represented by the Internet of Things, big data, cloud computing and artificial intelligence, is a typical representative of breakthrough technological innovation. At the same time, the traditional industries with the influence of the new generation of information technology tend to be integrated and developed, and the service-oriented manufacturing industry has become the development trend of the transformation and upgrading of traditional industries. Building and perfecting the industrial chain with the new generation of information technology, constructing regional technology innovation service platform, focusing on the development of supporting producer services and improving the development advantages of industrial clusters are important directions to give full play to the economic effects of the new generation of information technology and realize industrial transformation.

Finally, finance should play an important role in the process of disruptive technological innovation and industrial transformation. We need to improve the quality of financial supply and provide financial services to micro, small and medium-sized enterprises, innovate financing products and services, improve the financing efficiency of enterprises, effectively reduce hidden transaction costs and risks, and increase credit supply to micro, small and medium-sized enterprises. In addition, we should encourage financial institutions to provide services for industrial chain financing, financial leasing, and innovative investment in the process of serving the real economy.

8.4 New Measures for Chinese Industry Transformation

In recent years, China has issued a series of measures to adjust the industrial structure and promote industrial transformation and upgrading, but the effect is still not obvious. China's manufacturing industry is not driven by innovation, and China's high-tech field is still at the low end of the global value chain, which lags far behind developed countries. All these show that China has not completely solved the problem of industrial structure. The photovoltaic industry, once an emerging industry, has been strongly supported by the government. The photovoltaic industry, once an emerging industry, has received strong support from the government and various preferential policies given by the government have made photovoltaic companies in various places expand their scale. However, the rapid expansion of production capacity has led to a large amount of excess capacity in the short term. And China's photovoltaic industry is concentrated in the middle of the industrial chain, the upstream is limited by the price of foreign raw materials, the downstream is limited to the European market. Not only the photovoltaic industry, but other industries also have incomplete industrial chains, and enterprises lack their position in the industrial chain. Therefore, the adjustment of China's industrial transformation strategy has become a top priority.

8.4.1 Macro Measures of Industrial Transformation

The three revolution in the new industrial revolution, which are renewable resources, digital manufacture and the Internet information technology revolution, have actually become topics that cannot be avoided in the establishment of a new industrial system and industrial structure, and are the key to establishing a new industrial system and optimizing the industrial structure in China, which is also an aspect of industrial development that developed countries in Europe and the United States have been investing in and increasing policy support for recently. From the perspective of macro-measures, we should grasp several aspects.

8.4.1.1 Stimulating Market Vitality

China's existing industrial policies have too much regulation on the market and too little protection on the market itself. A market without sufficient competition is difficult to cultivate large enterprises with competitive strength in the global competition. In recent years, in some emerging industries that are less regulated by the government, some enterprises in China already have the ability to compete with multinational companies in international competition. For example, China's e-commerce industry has been started for more than 10 years, and the full market competition has given enterprises the vitality of competition. The leading enterprises of electronic commerce are basically private enterprises with distinctive characteristics. Jingdong mall has far more market share than Amazon in China and is gaining an edge in competition with traditional players such as Gome and Suning. E-commerce enterprises integrate upstream and downstream at the same time, driving the development of a series of industries such as production, logistics and warehousing. In China, large enterprises with the ability of industrial chain integration are more likely to emerge in fully competitive industries. Therefore, the government should continue to make efforts on how to establish and perfect the market mechanism, how to protect the full competition of the market, how to regulate the market and create a fair market environment.

8.4.1.2 Taking Strategic Emerging Industries to Achieve Technological Leapfrogging

We should organize research forces across society in the three fields of renewable energy, digital manufacturing and Internet information technology, and use new mechanisms to form strategic alliances between industries, universities and research institutes. In addition, innovation must have a certain strategic direction in order to realize the accumulation of knowledge and promote the upgrading of industrial and value chains. The key to the development of strategic emerging industries is to achieve technological breakthroughs, improve the independent innovation ability of enterprises, realize the upgrading of the industrial chain and value chain, and obtain the governance and control rights of the industrial chain and value chain. The knowledge spillover effect and industrial association effect brought by the innovation of strategic emerging industries can drive the evolution of China's entire industrial system and realize the transformation of industrial structure.

8.4.1.3 Maintaining the Openness to Achieve Interconnection with the Global Value Chain

In the context of globalization, it is impossible for China to realize the transformation to a new industrial system and structure in isolation. In the process of establishing a new industrial system, market space plays a decisive role in the technological space of industrial transformation. China should expand its market space through opening up to accommodate more products, and at the same time using the international market to realize the transformation from vertical labor division to horizontal labor division. China should adhere to the joint development of resources and projects with foreign companies, and make full use of the advantages of countries along the "Belt and Road" for cooperation. In the end, we need to realize that Chinese products can not only meet the needs of Chinese residents for a better life, but also meet the mid-to-high-end needs of foreign residents, change the product supply model that focused on quantity in the past and turn to a model that pays more attention to quality. In the global value chain, the peripheral relationship that relies on the core technology and core products of other countries for supporting has turned to the core relationship dominated by China. We should grasp high-end technologies of the global value chain through independent innovation, fully implement the strategy of innovation-driven development, and make innovation the strategic underpinning for building a modernized economy.

8.4.2 Micro Measures of Industrial Transformation

Cultivating a large number of leading large enterprises in the global industrial chain and value chain is not only necessary for the transformation of industrial structure, but also the key to enhancing the international competitiveness of Chinese industries. To this end, we need to have a good environment suitable for the survival and development of enterprises, but also need to provide sufficient impetus and support for the development of enterprises.

8.4.2.1 Guiding Large Enterprises to Carry Out Endogenous Changes and Cultivate Core Competitiveness

Chinese enterprises need to carry out endogenous transformation of enterprises in the uncertain operating environment of reality and future, which requires strategic transformation of enterprise development. The key to success may depend on two things. The first is the ability of enterprises to accurately grasp the external environment. With the rapid development of information technology and other high technologies, the external business environment accelerates change and enterprises are experiencing the transition from the era of scale competition to the era of speed competition. Therefore, in the process of reforming itself and realizing strategic transformation, enterprises should not only have a clear understanding of the external environment, but also enhance their ability of speed of understanding, that is, the ability to have immediate warning of environmental changes and to respond quickly to environmental changes.

Second, the key resources should match the external environment inside the enterprise. The change of enterprise external environment is the inducement of enterprise strategic transformation. Whether an enterprise can make a comprehensive grasp and rapid response to the change of environment or obtain core competitive advantage mainly depends on the specific knowledge of the enterprise. Therefore, the construction of enterprise internal specific knowledge is another important factor to ensure the success of strategic transformation.

8.4.2.2 Supporting Large Enterprises to Carry Out Mergers and Acquisitions

Industrial structure adjustment is an important way of industry adjustment of stock assets. Except the so-called evolution, the market method of stock asset adjustment is the purchase and integration of assets of similar enterprises in the industry or upstream and downstream enterprises in the industry chain by advantageous enterprises through mergers and acquisitions. At the same time, it is beneficial to exert synergistic effect. Improving low-efficiency assets to be well allocated by integrating resources, we can achieve better scale and scope effects, shorten the time to market for new products and achieve industrial chain. Therefore, for the current situation of China's industrial structure problems and serious structural imbalances, the government should support advantageous enterprises to merge and reorganize existing assets in other industries. It is important to give full play to the guiding role of the government, adopt multi-party cooperation, establish a multi-level financial market system and effectively enhance the financial support capability for mergers and acquisitions of advantageous enterprises. Through financial innovation, regulatory innovation and institutional innovation, we will establish a policy guarantee system that conforms to the laws of industrial and investment development for mergers and acquisitions of competitive enterprises from the source, and support them to develop into industry-leading enterprises.

8.4.2.3 Strengthening Knowledge Management for Long-Term Competitive Advantage

Knowledge is path-dependent, and the incremental knowledge of an enterprise is strictly dependent on the existing stock knowledge. If the stock knowledge of an enterprise creates competitive advantages for the enterprise, such advantages will be maintained with the generation of incremental knowledge in a stable environment, which is reflected in the continuity or self-enhancement of competitive advantages. Path dependence is closely related to historical dependence and cumulative learning. On the one hand, the reason why knowledge can become the source of continuous competitive advantage of enterprises is determined by the type and characteristics of knowledge stock owned by enterprises, that is, the combination of knowledge with intentionality and process, complexity and interrelation produce "cause ambiguity", which makes it difficult to observe and imitate. On the other hand, the path dependence and historical dependence keep and expand the competitive advantage. Therefore, in the case of efficient market, the competitive advantage of enterprises comes from the specific knowledge owned by enterprises. Knowledge not only earns Ricardo rent and Chamberlain rent for enterprises, but also creates "Schumpeter rent" for enterprises through continuous knowledge innovation through organizational learning. In addition, the sustainable competitive advantage of enterprises must depend on the knowledge protection system of enterprises, that is, enterprises provide a kind of "isolation mechanism" to protect the knowledge innovation of enterprises from being usurped or imitated by competitors.

8.4.3 The Role of Government in Industrial Transformation

The restructuring of industrial chain led by large enterprises will adjust and change the micro part of industrial structure, and ultimately promote the transformation and upgrading of industrial structure. The key to the success of such a market adjustment method also needs to clarify thc correct role positioning of the government in the transformation of industrial system and structure.

8.4.3.1 Maintainer of Market Competition Environment

In the market economy, the right role of the government in industrial development and structural adjustment should be to maintain a fair and just market competition environment, reduce the possibility of enterprises taking opportunistic actions, and promote the healthy development of the market. The government should do a good job in providing public services for economic development and industrial structure adjustment, and let the market play a better role in the optimization of industrial structure. The government should also provide economic and industrial policies against the background of global industrial development and market competition, but not rely on administrative power to "accelerate" industries and enterprises. In the current environment of economic integration, due to the global distribution of industrial chains and value chains, China's industries and enterprises are not only facing competition in one region or within a country, but also the competition in global industries and enterprises. The advantage is to gain control over the industrial chain and value chain in the competition of the global market, which requires the ability to achieve high efficiency and innovation in the complex global market.

The government should be the maintainer of a good market environment, and should create a market environment conducive to industrial chain competition and enterprise competition. At the same time, the government should design industrial regulation system and industrial policy system to promote industrial development and industrial structure transformation through market mechanism, rather than direct investment to promote industrial development and replace enterprises to achieve development. Only in a healthy and orderly market, through the market mechanism of resource allocation in the economic system to achieve superior bad discard can help enterprises obtain fair market opportunity, giving full play to the advantages of its ability to innovate, improve efficiency, the development ability, thereby giving impetus to the ascension of the whole industry chain and achieve the goal of industrial structure optimization.

8.4.3.2 Executor of Scientific Industrial Regulation

As the maintainer for the market fair, the government should respect the market's choices in the evolution of the industrial system and the transformation of the industrial structure, the collaborative innovation choices of leading enterprises in the industrial chain, and the governance of the collaborative relationship between enterprises in the industrial chain. But it does not mean that the government has completely withdrawn from economic development and industrial development, because the market sometimes fails, especially because enterprises pursue monopoly, which will lead to monopoly industries and monopoly technologies, making the market unable to carry out fair and effective competition, and ultimately not conducive to innovation and development. Finally, it leads to a decrease in the efficiency of resource allocation. China's industrial system and structure are unbalanced. Monopolized markets coexist with over-competitive markets. Some industries have been over-monopolized, such as petrochemical energy, electric power, telecommunications while some other industries are precisely over-competitive, with fierce price competition and sharp overcapacity. Both conditions are manifestations of low efficiency of resource allocation in China, which is closely related to the government's administrative approach to promote local GDP and industrial development, as well as China's failure to design industrial regulations and the government's failure to properly implement the anti-monopoly Law.

After more than 30 years of reform and development, China's industrial system and structure have faced a critical period of transformation. The demographic dividend and land dividend are gradually disappearing, the environmental pressure is increasing, and the global economic integration has made the distribution of the value chain more uneven. The extensive management methods of government comparisons are no longer suitable for the current actual situation. It is urgent to design a scientific and effective industrial regulation system to regulate the healthy evolution of new industrial systems and structures. The government should learn from the economic management methods of mature markets, including becoming a good enforcer of industrial regulations and laws. We should give priority to the development of high-tech, high-output, low-consumption, low-emission industries and enterprises, achieve the optimal allocation of resources to the maximum extent, maintain the balance of industrial ecology, promote the harmonious development of man and nature, and achieve the unification of economic development, population, resources and environment.

8.4.3.3 Cultivator of the Industrial Chain Control

As the leader of the industrial chain, large enterprises promote the adjustment and optimization of industrial structure through industrial correlation and collaborative innovation with related enterprises and industries. The key to success is to have such leading enterprises of the industrial chain and value chain, or what we call controllers and system integrators of the industrial chain. However, there is a serious lack of such large-scale system integrators which have control over the industrial chain and value chain in China. We also have some large enterprises, but they are basically cooperators in other countries' industrial chains and do not have control over them. Therefore, how to cultivate a large number of enterprises with dominant control or integration ability in different industrial chains, value chains and even supply chains is very important. What's more, the international competitiveness of an industry actually depends on whether the industry has outstanding large enterprises with influence and industrial chain control in the international market, such as Boeing, IBM, GE, Microsoft, Apple, etc. Their global competitiveness has demonstrated the international competitiveness of the United States in large aircraft manufacturing industry, computer software industry, mobile phone manufacturing industry and so on.

Therefore, as a developing country and a country that is developing in the global industrial competition, the government is cultivating large enterprises to become the controllers of the future industrial chain and value chain, which is actually cultivating the international competitiveness of China's industries. In the process of cultivating leading enterprises or integrators in the industrial chain and value chain, the government should play the role of "leader" rather than the role of "nanny". In the guiding process of cultivating large enterprises to become integrators with control over the industrial or value chain, the government needs to avoid the following aspects. First of all, it is necessary to avoid unilaterally promoting enterprises to pursue large-scale development, or forcibly assembling enterprise groups through administrative means such as blindly uniting, which seems to expand the scale of enterprises on the surface, but does not really play a role in the integration and reorganization of the industrial chain. Second, we need to avoid in the development of large enterprises in the process of growth, to rely on market competition to achieve excellence and inferiority. Finally, avoiding blind support backward technology, environmental protection, high consumption of inefficient, only production scale advantages of large enterprises, but to cultivate the innovation ability and competitive advantage of enterprises, encourage these enterprises along the industry chain and value chain control, become a leader controller, the industrial chain and value chain.

References

Clark, Colin. 1939. The Marginal Productivity of Capital. *Economic Record* 15 (3): 110–118.

Cui, Weijie. 2010. Current Situation, Problems and Countermeasures of International Market Development of Strategic Emerging Industries. *Intertrade* 10: 22–26.

Das, Dilip K. 2010. Contours of Deepening Financial Globalization in the Emerging Market Economies. *Global Journal of Emerging Market Economies* 2 (1): 45–67.

Du, Longzheng, Yanming Wang, and Shi Li. 2010. Research on the Industry Chain's Governance Structure and Basic Models. *China Industrial Economics* (3): 108–117.

Feng, Xiaoqing. 2019. Application of the Intellectual Property Right System and the Innovation-Based Development of Enterprises-Insight of the US Ban on Huawei in the Context of the Sino-US Trade Friction. *Renming Luntan· Xueshu Qianyan* (13): 32–42.

Feser, J. Edward., and Edward M. Bergman. 2000. National Industry Cluster Templates: A Framework for Applied Regional Cluster Analysis. *Regional Studies* 34 (1): 1–19.
Frederick, S., and Ronald Verlin Cassill. 2009. Industry Clusters and Global Value Chains: Analytical Frameworks to Study the New World of Textiles. *Journal of the Textile Institute* 100 (8): 668–681.
Freeman, Christopher and Luc Soete. 1982. The Economics of Industrial Innovation: Third Edition.
Freeman, Christopher. 2008. *Technology Policy and Economic Performance Lessons from Japan.* Southeast University Press.
Gao, Weikai. 2010. The Decisive Factors of the Technical Innovation of China Equipment Manufacturing. *Modern Economic Research* 4: 40–43.
Ge, Ying. 2009. Globalization and Industry Agglomeration in China. *World Development* 37 (3): 550–559.
Gong, Huiqun, Chao Huang, Shu-e Mei, and Weijun Zhong. 2013. The Development Path of Emerging Industries in Globalization and Its Enlightenment to China. *Science & Technology Progress and Policy* 30 (3): 51–54.
Guo, Tianjiao, and Guoqing Zou. 2020. Research on the Open Innovation Model of Strategic Emerging Industries and Countermeasures. *Economic Review Journal* (3): 102–107.
Ito, Tadashi, Richard E. Baldwin and Hitoshi Sato. 2015. The Smile Curve: Evolving Sources of Value Added in Manufacturing. *Proceedings of the Winter Academic Presentation of the Korean Society of International Economics*, 2015: 10.
Jiang, Guojun and Yifei Du. 2003. Research on the Pricing Range of Intermediate Products in Industrial Chain. *China Economist* (12): 21–23.
John, R. Hicks. 1999. John R. Hicks on the Economics of Population. *Population and Development Review* 25 (2): 345.
Joseph, A. Schumpeter. 2019. *Theory of Economic Development.* Beijing: The Chinese Overseas Publishing House.
Lall, Sanjaya, Manuel Albaladejo, and Jinkang Zhang. 2009. Mapping Fragmentation: Electronics and Automobiles in East Asia and Latin America. *Journal of Translation from Foreign Literature of Economics* (4): 29–45.
Li, Huai-liang. 2020. From Globalization Age to Global Community Age. *Modern Communication (Journal of Communication University of China)* 42 (6): 1–5.
Li, Huiru and Zhaowei Chen. 2018. The Influence of "The Belt and Road" Proposal on Regional Value Chain Construction of High-end Industry. *Journal of Hebei University of Economics and Business* 39 (4): 36–44, 65.
Li, Yan. 2019. Global Cyberspace Governance: State Actors and the China-US Cyber Relationship. *Contemporary International Relations* 39 (2): 105–124.
Liao, Rebecca. 2016. Dysfunction, Incentives, and Trade: Rehabilitating U.S.-China Cyber Relations. *Georgetown Journal of International Affairs* 17 (3): 38–46.
List, Friedrich, Sampson Lloyd and J. Shield Nicholson. 1916. *The National System of Political Economy*. Longmans, Green and Co.
Liu, Xue-lian and Song-mei Zhang. 2007. On the Contradiction of Globalization Process and China's Action. *Jilin University Journal Social Sciences Edition* 60 (2): 119–126,221–222.
Ma, Shuang. 2020. Huawei's International Expansion of 5G against the Background of Sino-U.S. Competition: A Political Risk Analysis. *Contemporary Asia-Pacific Studies* (1): 4–29,158.
Mandelbaum, K. 1948. The Industrialization of Backward Areas. *International Affairs* 24 (2): 259.
Munir, A. Kamal., and Nelson Phillips. 2002. The Concept of Industry and the Case of Radical Technological Change. *The Journal of High Technology Management Research* 13 (2): 279–297.
Paul, M. Romer. 2009. Increasing Returns and Long-Run Growth. *Innovation and Entrepreneurship*, 103–138
Peter, F. Drucker. 1985. The Discipline of Innovation. *Harvard Business Review* 63 (3): 67–72.
Peter, R. Mueser. 1987. Identifying Technical Innovation: IEEE Transactions on Engineering Management. *The Journal of Science Policy and Research Management* 2 (4): 496.

Ren, Baoping, and Yinxing Hong. 2005. An Approach of Improvement of the Economic Effectiveness in New—Type Industrialization: Analysis of the Visual Angle of a Kind of Industry Chain. *Journal of Northwest University (Philosophy and Social Sciences Edition)* 35 (1): 47–54.

Rosenstein-Rodan, Paul. 1943. Problems of Industrialization of Eastern and South-Eastern Europe. *The Economic Journal* 53: 202–211.

Scott, Allen J. 2006. The Changing Global Geography of Low-Technology, Labor-Intensive Industry: Clothing, Footwear, and Furniture. *World Development* 34 (9): 1517–1536.

Staley, Eugene. 1939. World Economy in Transition; Technology vs. Politics, Laissez Faire vs. Planning, Power vs. Welfare. *Council on Foreign Relations.*

Sun, Yanhong. 2019. Exploring the New Position of the Government's Economic Role—An Analysis of the UK's Industrial Strategy Since the Outbreak of the International Financial Crisis. *Chinese Journal of European Studies* 37 (1): 68–90, 6–7.

Tang, Zhiwei, Yuxuan Li, and Longpeng Zhang. 2021. Identification Method and Breakthrough Path of "Neck-Jamming" Technologies under the Background of Sino-US Trade Friction: A Case of the Electronic Information. *Science & Technology Progress and Policy* 38 (1): 1–9.

Wang, Songji, Gu., Lei, Su., and Xiaoqing. 2019. The "Belt and Road" Initative Leads to New Globalization: Changing Pattern and Construction Task. *Journal of Northwest University (philosophy and Social Sciences Edition)* 49 (2): 148–154.

Wang, Xiaowen, Kunpeng Ni, and Xiaojin Ma. 2015. Research on the Relationship Between Emerging Industry Agglomeration and Industrial Scale from a Regional Perspective. *Guizhou Social Sciences* (2): 110–118.

Yu., Wang, Chunhui Gan, and Wei Wang. 2013. The Analysis of Demand Motive for the Evolution of Industrial Structure: Study Based on Non-competitive Input-Output Model. *Journal of Finance and Economics* (10): 60–75.

Wei, Long, and Lei Wang. 2016. From Embedding Global Value Chains to Leading Regional Value Chains: The Economic Feasibility Analysis of the "One Belt and One Road" Strategy. *Journal of International Trade* (5): 104–115.

William, J. Baumol., and K. Alvin. 1970. Input Choices and Rate-of-Return Regulation: An Overview of the Discussion. *Bell Journal of Economics and Management Science* 1 (2): 162.

Wong, Poh Kam, and J.H. Mathews. 2005. Competitive Strategies of Asian High-tech Firms: The Challenge of Late-industrialization. *International Journal of Technology Management: The Journal of the Technology Management of Technology, Engineering Management, Technology Policy and Strategy* 29 (1): 1–5.

Yu, Dianfan, Chunhui Gan, and Ruogu Zheng. 2011. Industrial Linkages in China-An Empirical Research Based on Structural Decomposition Technique of Input-Output Model. *China Industrial Economics* (11): 5–15.

Zhao, Quanfu, and Zongwei Liu. 2016. Industrial Transformation Strategy for the Manufacturing Sector of China in the Tide of Industry 4.0. *Forum on Science and Technology in China* (1): 58–62.

Zhou, Alvin Y. 2018. Society & the Internet: How Networks of Information and Communication Are Changing Our Lives. *Information, Communication & Society* 21 (12): 1769–1771.

The manufacturer's authorised representative in the EU is Springer Nature Customer Service Centre GmbH, Europaplatz 3, 69115 Heidelberg, Germany. If you have any concerns regarding our products, please contact ProductSafety@springernature.com

Printed and bound by CPI Group (UK) Ltd, Croydon, CR0 4YY
30/09/2026
02225242-0002